Chimpanzees of the Lakeshore

Chimpanzees are humanity's closest living relations, and are of enduring interest to a range of sciences, from anthropology to zoology. In the West, many know of the pioneering work of Jane Goodall, whose studies of these apes at Gombe in Tanzania are justly famous. Less well-known, but equally important, are the studies carried out by Toshisada Nishida on the eastern shore of Lake Tanganyika. Comparison between the two sites yields both notable similarities and startling contrasts. Nishida has written a comprehensive synthesis of his work on the behaviour and ecology of the chimpanzees of the Mahale Mountains. With topics ranging from individual development to population-specific behavioural patterns, it reveals the complexity of social life, from male struggles for dominant status to female travails in raising offspring. Richly illustrated, the author blends anecdotes with powerful data to explore the fascinating world of the chimpanzees of the lakeshore.

TOSHISADA NISHIDA (1941–2011) was Executive Director of the Japan Monkey Centre and Editor-in-Chief of the journal *Primates*. He conducted pioneering field studies into the behaviour and ecology of wild chimpanzees for more than 45 years.

Chimpanzees of the Lakeshore

Natural History and Culture at Mahale

TOSHISADA NISHIDA
Japan Monkey Centre

Shaftesbury Road, Cambridge CB2 8EA, United Kingdom

One Liberty Plaza, 20th Floor, New York, NY 10006, USA

477 Williamstown Road, Port Melbourne, VIC 3207, Australia

314–321, 3rd Floor, Plot 3, Splendor Forum, Jasola District Centre, New Delhi – 110025, India

103 Penang Road, #05–06/07, Visioncrest Commercial, Singapore 238467

Cambridge University Press is part of Cambridge University Press & Assessment, a department of the University of Cambridge.

We share the University's mission to contribute to society through the pursuit of education, learning and research at the highest international levels of excellence.

www.cambridge.org
Information on this title: www.cambridge.org/9781107601789

First published 2012

A catalogue record for this publication is available from the British Library

Library of Congress Cataloging-in-Publication data
Nishida, Toshisada, 1941–
The chimpanzees of Mahale : natural history and local culture / Toshisada Nishida.
p. cm.
Includes bibliographical references and index.
ISBN 978-1-107-01578-4 (hardback) – ISBN 978-1-107-60178-9 (paperback)
1. Chimpanzees – Behavior – Tanzania – Mahale Mountains National Park.
2. Chimpanzees – Research – Tanzania – Mahale Mountains National Park.
I. Title.
QL737.P96N565 2012
599.88509678′28–dc23

2011033621

ISBN 978-1-107-01578-4 Hardback
ISBN 978-1-107-60178-9 Paperback

Contents

Colour plates are located between pages 236 and 237.

Foreword

The book you hold in your hands, with its fine photographs and exquisite descriptions of chimpanzee behaviour by one of the world's greatest experts, would have been unthinkable half a century ago. We have come such a long way in our knowledge of chimpanzees, and the discoveries have reached us in such a gradual and cumulative fashion, that we hardly realise how little we used to know about our nearest relatives.

At the time, chimpanzees did not yet occupy the special place in our thinking about human evolution that they do today. Strangely enough, science looked at baboons as the best model of our ancestors, as baboons, too, had descended from the trees to become savanna-dwellers. These rambunctious monkeys, however, are quite far removed from us. For one thing, they have tails. Apes and humans belong to a small superfamily within the primate order, known as the hominoids, which are marked by flat chests, relatively long arms, large body size, superior intelligence, and the absence of a tail. Apart from humans and chimpanzees, living members of the superfamily include only gorillas, bonobos, orangutans, and gibbons.

Interest in apes started relatively late. Early primatologists had seen them travel through the trees, eating fruits at their leisure, but rarely did they notice anything of interest in their behaviour. This was partly due to low visibility in the forest and the apes' wariness of people. They disappeared as soon as they heard or saw observers approach.

The study of chimpanzee behaviour began in earnest only in the early 1960s, near the shore of Lake Tanganyika, in Tanzania. Two camps were set up, one by Western scientists at Gombe Stream and one 170 km to the south, at the foot of the Mahale Mountains, by Japanese scientists. The author of this book, Professor Toshisada Nishida of Kyoto University, started the latter camp in 1965. He was still a graduate student at the time but has worked in Mahale ever since, up to and including his retirement.

One of Dr Nishida's very first discoveries was truly groundbreaking. While science still described chimpanzees as sort of peaceful vegetarians that roamed the forest without any need for social bonds – not unlike Rousseau's noble savages, who did not need each other to survive – Dr Nishida noticed that chimpanzees live a communal life with clear territorial boundaries and perhaps even hostility between neighbouring unit-groups. This was not an easy discovery, because chimpanzees are often encountered alone or in small groups in the forest, so one can determine community relations only if one knows all individuals and keeps careful track of their travels. Dr Nishida's discovery not only upset Western notions of chimpanzees as individualists, but also the ideas of Dr Nishida's own teachers, who expected chimpanzees to live, like humans, in nuclear family-like arrangements. Debate about what to expect must have been rather heated, because when Dr Nishida's teacher, Professor Junichiro Itani, arrived in Kigoma, the student couldn't wait for their actual reunion, and shouted from aboard the steamship *Liemba*: 'There is no familoid in the chimpanzee society.' Professor Itani shouted back: 'That can't be true!'

We have learned much about chimpanzees since then, such as that they hunt and eat meat; that they raid their neighbour's territory and occasionally kill each other; that they use a complex set of tools, which differs from group to group; that they medicate themselves with plants; that males engage in power politics while competing over status and females; and so on. The list of discoveries is impressive, and the Mahale field site has been absolutely central in furnishing the evidence. From the start, the approach followed at Mahale has been that of the grand teacher of both Dr Nishida and Professor Itani, Professor Kinji Imanishi, who urged his students to identify individuals and to follow them over time. Not just for weeks or months, as previous studies had done, but for years and years, so that one began to understand the kinship relations within the group.

With a species that breeds as slowly and is as long-lived as the chimpanzee, one needs to follow individuals for a long time to know whether or not two adult males are brothers, or how many offspring a female raises during her lifetime. Before scientists learned to analyse DNA evidence, the only way to know much about genetic relatedness was a long-term project like the one Dr Nishida set up.

The first challenge was obviously to get to see the chimpanzees on a regular basis, identify them by their facial and bodily characteristics and get them so used to human presence that they would display their natural behaviour. At the time, habituation was typically done by means of food provisioning. Dr Nishida first tried it with sugarcane, until he found out that bananas worked better. He developed a 'mobile provisioning' technique in which scientists would announce their presence to distant chimpanzees by imitating the species-typical hooting calls, after which the chimps would approach and obtain some food. This way, their normal roaming patterns remained intact, as they never attached themselves to a fixed feeding site. After the chimps had fed, the investigators would simply follow them for the rest of the day.

Some scientists have criticised food provisioning as a technique that may make chimpanzees more aggressive. This assumption was used as an argument against reports of lethal intergroup warfare in the chimpanzees of Gombe and Mahale, reports which stimulated much debate about the aggressive nature of our own species. If chimpanzees kill each other the way we do, so the argument went, we probably inherited our territorial tendencies from the ancestor we share with chimpanzees. There are many opponents of this view, who prefer to blame the violence among wild chimpanzees on food provisioning. However, while provisioning in Mahale ended in 1987, the aggressive behaviour of the chimpanzees hardly changed. Moreover, there are now reports of the same violent behaviour from field sites where researchers *never* provisioned any chimpanzees with food. For this reason, there is little doubt among experts that chimpanzees are just naturally violent.

Dr Nishida was a most dedicated scientist who, in his early career, spent years, and later many months per year, under relatively primitive circumstances, without running water or electricity, at Kasoje at the northern foot of the Mahale Mountains. As a result, he knew all chimpanzees in several groups. By 'knowing' I mean that he observed them as infants, saw them grow up as juveniles, followed them through their prime and into old age.

In one of the first studies of a power take over in wild chimpanzees, Dr Nishida observed 'allegiance-fickleness' in an old male, who was past the age of becoming alpha but who carved out a key position by regularly switching sides in alliances with two adult males competing for dominance, so that he achieved maximum social influence. Many other observations of power politics have followed, including an alpha male who used meat to 'bribe' others to support him. The overall conclusion has been that chimpanzees apply great strategic intelligence to their social relations.

The ecology of the species holds special fascination. Over the years, Dr Nishida tasted about 100 species of plants or fruits that chimpanzees eat at Mahale in order to get an idea of how the apes perceived these foods. A major advance in the study of cultural habits came when Dr Nishida discovered that chimpanzees consume *Aspilia* leaves, which they do very slowly, mostly in the morning, swallowing the leaves without chewing them. This aberrant ingestion seemed unlikely to be done for nutritional reasons. Together with Professor Richard Wrangham, the first Western primatologist to set foot in Mahale, Dr Nishida published his observations of potentially medicinal use of plants by wild chimpanzees, thus founding the new field of zoopharmacognosy.

Over the years the Mahale team has welcomed many other collaborators, and as a result has an extremely rich and diverse publication record. Dr Nishida was one of the most respected scientists our field has ever produced, and in 2008 was rightly presented, along with Dr Jane Goodall, with the prestigious Leakey Prize, which recognises accomplishments in human evolutionary science.

In March 2004 I attended Dr Nishida's retirement lecture at Kyoto University. The lecture was riveting, especially given all the historical details of how our knowledge has grown over the years and the role of Japanese scientists. For me, extra pleasure came from having visited Mahale just the year before. I had found it an enchanting place where the chimpanzees are so well-habituated that it is not hard to see how it could have produced so many historic discoveries. *Chimpanzees of the Lakeshore* reviews all of these discoveries, offering myriad behavioural details, which readers can be sure the author saw himself, with his own eyes, describing them for the first time for a large public, in his own words.

Frans de Waal

Dr Frans B. M. de Waal is C. H. Candler Professor at the Psychology Department and Director of the Living Links Center, both at Emory University in Atlanta, USA. He is internationally known for numerous popular books on primate (and human) behaviour, including *Chimpanzee Politics* (1982) and *Our Inner Ape* (2005).

Preface

Forty-five years ago I first set foot in Kasoje, a treasure trove for chimpanzee research set in the Mahale Mountains and bordered by Lake Tanganyika. I started out as a graduate student at the age of 24. Time has passed since then, and it still passes; most of today's active researchers were born after my research began. When I was an undergraduate student, I was interested in the exploration of *terra incognita* such as the Amazon, Borneo, Sumatra, or Africa. It was the early 1960s, only 15 years after the end of the Second World War, and Japan had not yet achieved a level of foreign exchange that allowed ordinary people to travel abroad. At that time, three books appeared on gorilla expeditions, written by Kinji Imanishi, Junichiro Itani, and Masao Kawai. These books not only filled me with interest in the great apes but also made me realise that if I studied chimpanzees, I could go to Africa! Although I was a zoology student, I only vaguely imagined before reading these books that I might study the ecology of animals. Fortunately, the graduate course of Physical Anthropology in the Zoology department, with Imanishi as the first professor, was established in 1962, as if it were prepared just for me. Therefore, I was happy to enter the course with my colleagues, Takayoshi Kano and Kohsei Izawa. I thought I would study chimpanzees for three years or so and then change my research target to human beings and their traditional lives as hunter-gatherers.

Once I began to study chimpanzees, I soon realised they were not creatures that could be understood in a few years of study. We encountered new discoveries in behaviour just as surely as a new year arrives and the old year departs. Every year I found new dietary items, new behavioural patterns, new personalities, and new relationships. Individuals who appeared by birth or immigration have fascinated me by their doing something new. My first observations of chimpanzees encountering an unexpected animal were recorded for a crocodile in 1983, a lion in 1989, and a freshly dead leopard carcass in 1999. Similarly, cannibalism and ostracism were first observed more than two decades after the start of my research. The apes' behaviour is so rich in variety that, no matter how many years I observe them, I will never grow bored. Every year a new student comes to Mahale and surprises me by reporting a behavioural pattern that I have never seen before. Such discoveries happen because of the flexible behaviour of chimpanzees and the wonderfully diverse environment of Mahale, which is rich in biodiversity and landscape variation. This is why I have continued to conduct research for so long.

Chimpanzees have a diverse behavioural repertoire, rich in versatility, which allows them to adapt to complex natural, social, and demographic conditions. People have been searching for a 'missing link' in remote places in the Himalayas, China, and even North America. They have searched far and wide, but the answer was right before their eyes: chimpanzees are the missing link!

I wrote this book to introduce laypersons and students to the wondrous behaviour of the chimpanzee. I want to present the reader with a myriad of marvellous examples illustrating how a chimpanzee's behaviour will adapt itself to meet any circumstance. The basis of the book comes from my previous books that were written in Japanese (Nishida 1973a, 1981, 1994, 1999, 2008b) but a great deal of new data are added. My greatest wish is, through the rich use of photos and illustrations, to bring the reader closer to the actual world of the chimpanzee. I would be thrilled to learn that this book has helped readers to share my feelings about how closely humans and chimpanzees are actually linked.

This is my personal record of the chimpanzees of Mahale. As many of my colleagues have pointed out, chimpanzees of different study sites show different behavioural patterns, although, of course, they also have many in common. Just as a cultural anthropologist writes on the ethnography of a tribal society under study, I have written on the ethnography of the Mahale chimpanzees. Therefore, no one should assume that anything I have written about Mahale is applicable to the chimpanzees of other study sites.

Of course, humans and chimpanzees have many differences, and this is what makes it all so interesting – this is what breeds insight. Humankind's coexistence with different species brings to our lives great benefits and delight.

Moreover, I want my feelings on the plight of the chimpanzee to hit home, increasing the awareness that humankind has no right to monopolise the Earth, to waste its resources, to endanger masses of its living creatures. I hate to imagine living in a world where there is no dragonfly or no butterfly.

Acknowledgements

I am greatly indebted to eight persons, in particular, for the publication of this book. First, Bill McGrew not only commented on my earlier drafts, but also took over the whole work after my health declined. He synthesised all of the chapters, and even negotiated its publication with Cambridge University Press. John Mitani gave me critical but constructive comments on my earlier drafts, which, along with Bill's comments, improved the book considerably. Frans de Waal wrote an extraordinary Foreword for me. Linda Marchant, John, and Bill generously revised the English of two chapters each of the draft. Ron Read of Kurdyla and Associates also revised the English for six other chapters. Agumi Inaba performed miscellaneous editorial work, including digitising my photographs and figures to computer files. Kazuhiko Hosaka and Michio Nakamura assisted as my surrogates after I could not continue correspondence with Bill and the publisher.

This book is based on my long-term research of Mahale chimpanzees, for which I owe thanks to many people and institutions. First, I must thank the late Kinji Imanishi and the late Junichiro Itani, my two great mentors, for their pioneering theoretical and field study, which enabled us to initiate the Mahale Mountains Chimpanzee Research Project. I also thank Takayoshi Kano and Kohsei Izawa, who assisted in the initial study period, for their continuous support. Kano and I cooperated to promote a joint project on chimpanzees and bonobos throughout the 1980s. My colleagues for the longest time, the late Kenji Kawanaka and the late Shigeo Uehara, contributed greatly to the establishment and continuation of the long-term research. Without their support, this project would have withered on the vine. More recently, Kazuhiko Hosaka and Michio Nakamura also have made every effort to help continue this project.

I thank my many colleagues who have studied Mahale chimpanzees for their cooperation in all possible ways, including research, conservation, and management of the camp. A list of these valued comrades runs long: Junichiro Itani, Kenji Kawanaka, Makoto and Hideko Kakeya, Akio Mori, Shigeo Uehara, Bill McGrew, Caroline Tutin, Koshi Norikoshi, Masato Kawabata, Yukio Takahata, Mariko Hiraiwa-Hasegawa, Toshikazu Hasegawa, Hitoshige Hayaki, Anthony Collins, Masayasu Mori, Hiroyuki Takasaki, Richard and Jennifer Byrne, Michael Huffman, Ken'ichi Masui, Kevin Hunt, Rogath Olomi, Takahiro Tsukahara, Satoshi Kobayashi, Miya Hamai, Linda Turner, John Mitani, Tamotsu Aso, Miho Nakamura, Kazuhiko Hosaka, Kozo Yoshida, Akiko Matsumoto-Oda, Hiroko Yoshida, Koichi Koshimizu, Hajime Ohigashi, Mikio Kaji, Fumio Fukuda, Michio Nakamura, Mitsue Matsuya, Hiroshi Ihobe, Noriko Itoh, Hitoshi Sasaki, Linda Marchant, Hideo Nigi, Tetsuya Sakamaki, Nadia Corp, Koichiro Zamma, Christophe Boesch, James Wakibara, Takahisa Matsusaka, Nobuyuki Kutsukake, Shiho Fujita, Masaki Shimada, Hitonaru Nishie, Eiji Inoue, Mariko Fujimoto, Shunkichi Hanamura, Mieko Kiyono, Takanori Kooriyama, and Agumi Inaba.

I could not have contributed to science without the dedicated help of our Tongwe field assistants, cooks, boat drivers, house builders, and trail cutters: Ramadhani Nyundo, the late Omari Kabule, the late Juma Kahaso, Sadi Katensi, Mkoli Saidi, the late Issa Kapama Ally, Samola or Mosi Hamisi, the late Athmani Katumba, the late Almasi Kasulamemba, the late Kabukula Kasulamemba, Mohamedi Seifu, Haruna Sobongo, the late Haruna Huseni Kabombwe, Yassini Kiyoya, the late Ramadhani Kabilambe, Ramadhani Kasakampe, Jumanne Katensi, Kijanga or Rashidi Hawazi, Moshi Bunengwa, the late Mtunda (Mwami) Hawazi, the late Hamisi Bunengwa, Rashidi Kitopeni, Hamisi Katinkila, the late Luhembe Ismaili, Kabumbe Athumani, Bunde Athumani, Mosi Matumla, Mwami Rashidi, and many others. I thank especially Ramadhani Nyundo for his unparalleled expertise on Tongwe ethnotaxonomy of plants and animals as well as on the observation of chimpanzees.

I owe the village people of Kasoje deep thanks for their warm hospitality, in particular, Saidi Sobongo who accepted me as his son and called me 'mwanangu', his wives Wantendele and Binti Sudi, his father mwami Sobongo, and his uncle Bunengwa and his family. Saidi gave up his own room to me for my living space and storehouse during my first three years at Kasoje.

I owe deep gratitude, for their assistance and encouragement, to many people who were residents of Dar es Salaam during various periods, in particular: Kozo Tomita, Toshimichi Nemoto, Asami Kanayama,

Eiko Kimura, and Koichi Kobayashi. I obtained warm assistance from the diplomats of the Embassy of Japan and am indebted to Nobuyuki Nakashima, Yasuhiro Inagawa, Yuriko Suzuki, Yasushi Kurokochi, and Keitaro Sato for assistance in the JICA project and small-scale ODA.

In spite of their short-term visits, some colleagues such as Richard Wrangham, Hidemi Ishida, David Bygott, Barbara Smuts, Craig Stanford, Juichi Yamagiwa, and Frans de Waal stimulated me to think twice about my research from various viewpoints. Bill McGrew visited Mahale several times from 1974, and was the first Western scientist to conduct chimpanzee research at Mahale. Since then, he not only has constantly stimulated me to find behavioural patterns new to science, but also kept me informed of new developments at other study sites. John Mitani studied Mahale chimpanzees from 1989 to 1994. When the maintenance of the camp was difficult, he spent most of the time alone, making every effort to keep the chimpanzee database at a high level and to manage the camp effectively. His research on pant-hoots was so far the only contribution to the study of vocalisation at Mahale. Richard Wrangham was the first Western primatologist to visit Mahale in 1971. He told me a lot about the Gombe chimpanzees and his new finding of chimpanzees' leaf-swallowing behaviour. He has constantly stimulated me to reconsider chimpanzee behaviour from a sociobiological viewpoint.

I also express special thanks to Jane Goodall and Geza Teleki for alerting me to the dire situation of chimpanzee conservation in the 1980s. One of the results of this discussion is the continuous publication of *Pan Africa News* for research and conservation (http://mahale.web.infoseek.co.jp).

I am extremely grateful to Agumi Inaba (2001–present), Chisa Tokimatsu (1988–2003), Yoshiko Endo (1977–1988), and Nobuko Fukui (1972–1974) for their dedicated compilation of data and secretarial work. Naomi Miyamaoto also helped me to compile some important data on social relationships.

I express my deep gratitude to the Tanzania Commission for Science and Technology, Tanzania Wildlife Research Institute, and Tanzania National Parks for permission to conduct research and to the University of Dar es Salaam and the Wildlife Division of the Ministry of Natural Resources and Tourism for cooperation in research and conservation. I am extremely grateful to Professor Hosea Y Kayumbo for his long-time friendship and support. I thank Kapepwa Tambila, Costa Mlay, Gerald Bigurube, Charles Mulingwa, George Sabuni, Erasmus Tarimo, and Edeus Massawe for cooperation and encouragement. I also thank the Mahale

Mountains National Park and the Mahale Mountains Wildlife Research Centre for permission to do research and for logistic support.

The series of research on which this book is based has been supported financially by Grants-in-Aid for Basic Scientific Research of the Ministry of Education, Culture, Sports, Science and Technology (#03041046, 07041138, 12375003, 16255007, and 19255008), the Japan International Cooperation Agency, the Japan Society for the Promotion of Science, the Scholarship of Takenaka Engineering Firm, the Wenner-Gren Foundation for Anthropological Research, the Leakey Foundation, and the Global Environment Research Fund of the Ministry of Environment (F-061).

I am indebted to my great friends, Kenzo Itoh and Takashi Ichinose for their encouragement. Finally, but not least, I thank my family, my father Toshiharu, mother Taiko, wife Haruko, daughter Ikuko, son Toshimichi, brother Kiyoharu, and sister Yoko for their continuous support and encouragement. My wife and two children, in particular, shared my desire to see the completion of this book while caring for me after my condition worsened.

POST-SCRIPT TO ACKNOWLEDGEMENTS

Professor Toshisada Nishida ('Toshi' to his Western friends) died on 7 June 2011, aged 70 years, after a long battle with cancer. He died having seen the final version of the text of this book, and the final selection of photographs to illustrate it. He was most grateful to all of the people at Cambridge University Press, especially Martin Griffiths, who worked so hard to move the book along so quickly in the step-by-step process of publication. I, too, thank Amanda Friend and Jacob Negrey, who stepped in on short notice to help with the proof-readings.

WCM

Introduction

ORIGINS OF JAPANESE PRIMATOLOGY AND KINJI IMANISHI

My first mentor, Kinji Imanishi (1902–1992), was the founder of Japanese primatology. He was a bio-social anthropologist as well as ecologist, zoologist, entomologist, Himalayan mountaineer, explorer, and philosopher. His primary interest was the structure of the biological world, including human society (Imanishi 2002; Japanese original published in 1941).

From 1932 to 1942 he made several geographical and anthropological expeditions to Sakhalin, northern Korea, Mongolia, and northeastern China. From 1944 to 1945 he established the Seihoku (Northwestern) Research Institute at Choukakou and studied the ecology of pastoralists and their livestock in Mongolia. He returned to Japan at 1946 after the end of the Second World War in 1945 (Saitoh 1989). Finding no funds to conduct research outside of Japan, he began a study of the society of free-ranging horses indigenous to Japan at Toimisaki Point, Miyazaki Prefecture, in 1948. He identified and named each horse and investigated grouping patterns and social interactions among them.

One day in November 1948, when Imanishi's students, Shunzo Kawamura (1924–1999) and Junichiro Itani (1926–2001), were observing horses, they noticed that wild Japanese macaques in the distance were travelling in a neat procession. The beautiful line created by the monkeys' procession impressed them. After a month, Imanishi and the students visited Kohshima Island to look into the possibility of studying Japanese macaques. After this survey, Imanishi decided they should begin to observe macaques, leaving behind the horse research (Nishida 2009). Of course, he had already read Carpenter's pioneering primate studies,[1] and could predict that an investigation of monkeys would be interesting. Fortune knocks on the door of the person who is prepared for it.

Imanishi and his students successfully conducted research on the ecology, social structure, and behaviour of Japanese macaques through the methods of provisioning, individual identification, long-term research, and observation of many troops throughout Japan (Frisch 1959). During the macaque studies, Imanishi's interest was directed more to biological anthropology, as well as to comparative animal sociology, and in 1951 he wrote a book titled *Prehuman Societies*.

QUEST FOR THE MISSING LINK

In 1958 Imanishi organised the Japan Monkey Centre Gorilla Expedition, sponsored by the Nagoya Railway Company. The purpose of this research was to investigate the origin of the human family (Imanishi 1958a).

There are four types of great ape: chimpanzee, bonobo (bilia or pygmy chimpanzee), gorilla, and orangutan. These creatures are the closest living relatives to humanity. According to the traditional taxonomic nomenclature, they belonged to the same superfamily as humans (Hominoidea), but were classified into a different family, Pongidae, with only humans belonging to Hominidae.

Which ape should be selected for the research? Imanishi first considered orangutans because they were least known to the academic world. These Asian great apes intrigued Imanishi, who was always thinking of the 'first ascent' as an alpinist. However, he did not pursue this idea because, due to their arboreal habits, orangutans would be difficult to habituate through provisioning. Among African apes, virtually nothing was known at that time about the pygmy chimpanzees.

Gorillas were selected as Imanishi's first target, as he knew where they could be found and because gorillas were more terrestrial than chimpanzees. Hence he thought gorillas would be easier to habituate by provisioning than chimpanzees. On the basis of short preliminary surveys by two young women, Rosalind Osborn (in 1956–1957) working under the supervision of Louis Leakey, and Jill Donisthorpe working under the supervision of Raymond Dart (in 1957) (Sussman 2007), Imanishi reasoned that gorillas, and probably great apes in general, formed 'familoid' groups. He considered these to be a social unit of the

[1] See Carpenter (1964) for an anthology of his primate studies.

'species society'. On the other hand, from the ethnographic and cultural anthropological literature, he extracted the key factors that were common to human families worldwide. These are: (1) incest taboo; (2) exogamy; (3) division of labour between the sexes; and (4) existence of a 'community'. By 'community' he meant that social units do not exist independently but are integrated into a higher unit (Imanishi 1958b). This is the reason that I do not use the term 'community' for the social unit of chimpanzees or bonobos. Instead, I employ it in the manner Imanishi originally intended. Prior research suggested that gorillas had a family-like social unit, i.e. one male–multi-female group, but the relationships between social units were unknown.

The gorilla expedition lasted for three years (1958–1960) and clarified the social structure of the gorillas to some extent, but the political instability in what was then the Belgian Congo prevented the continuation of research. At the same time, George Schaller was initiating another study of gorillas in the Virunga Volcanoes region, which was to lead to Dian Fossey's well-known research there. Given these events, Imanishi changed his target of study from gorillas to chimpanzees. He apparently thought that both possessed a similar social structure, and that they would be interchangeable for research purposes. Gorillas and chimpanzees were believed to be sister species and equally related to humans at that time, and as a consequence, they were considered together to be our closest living relatives.

The start of the first Japanese ape expeditions coincided with a great wave of international interest in the great apes living in their natural environment. In the late 1950s, Adriaan Kortlandt (1962) started his research on chimpanzees at Beni in the Congo, while George Schaller (1963) started his mountain gorilla study in the Virungas. In 1960 Jane Goodall (1963) began her long-term study of chimpanzees at Gombe, Tanzania, and in 1963 Vernon Reynolds (Reynolds & Reynolds 1965) followed suit in the Budongo Forest, Uganda.

NO FAMILOID EXISTS

After the gorilla expedition, in 1961 Kinji Imanishi organised the first Kyoto University African Primate Expedition (KUAPE). He collected funds from many private companies, as well as from the Japanese Ministry of Education, Science, and Culture. He was allowed by the Vice-Chancellor to use Kyoto University for tax-free donations for overseas scientific research. From his experiences of alpine climbing and expeditions, he was an expert organiser.

I joined the Kyoto research team in 1965. When I did not find a 'familoid' system in chimpanzee society after successfully provisioning them in 1966 (see below), I reported this important information to Dr Itani, who had just arrived at Kigoma, Tanzania, from Japan. Itani was my second mentor after Dr Imanishi's retirement from Kyoto University in March 1965.

By 1970 I had concluded that chimpanzees have a unit-group as the social unit, which is a multi-male, multi-female group ranging from about 27 to 80 in size, and which usually splits up into temporary subgroups or parties. I also had discovered that relationships between unit-groups are antagonistic, that adult males are the core of the unit-group. Males are bonded with each other more closely than to members of any other age or sex class, and females, not males, transfer to other unit-groups (Nishida 1968; Nishida & Kawanaka 1972).

These social and behavioural characteristics differed remarkably from those of gorillas (Schaller 1963; Fossey 1970). This caused me immense difficulty when attempting to reconstruct the evolution of the human family. I was at a loss as to how to reconstruct the society of the last common ancestor (LCA), given the radically different social structures of chimpanzees and gorillas. At the time, these two species were believed to be equally related to humans, and it appeared impossible to unravel how human behaviour emerged from these two very different patterns.

After joining the Department of Anthropology at the University of Tokyo in 1969, I felt obliged to use the study of chimpanzees to ask questions about human origins. As I continued to struggle with these problems, two lines of new evidence emerged. The first involved my work with the pygmy chimpanzee's social structure. In February 1972 I did preliminary research on pygmy chimpanzees around the Lac Tumba region of the Equatorial region of Zaire (Nishida 1972a). The next year Takayoshi Kano and I visited the Salonga National Park, but we could not find a good site for a long-term study. Kano continued this general survey. He found and established a permanent field station at Wamba (Kano 1979), after travelling 2000 km around the country on a bicycle. Kano and his colleagues discovered that the bonobo's social structure is generally similar to that of the chimpanzee. Both live in multi-male, multi-female social units; unit-groups avoid each another; and females, not males, disperse from natal groups at adolescence. Despite these similarities, the two species also display some important social differences (Kano 1980; Kuroda 1979, 1980), such as female dominance over males in some contexts, but the similarity in social structure between them was confirmed. It was important to know that the phylogenetically closest species share a similar social structure. This finding was necessary to reconstruct the evolution of human society.

A second major discovery also helped to clarify matters. Morris Goodman had asserted as early as 1960 that the African apes were more closely related to man than were orangutans (Goodman 1961). His assertion was based on immunological comparisons between humans and apes, and he concluded that humans must be classed with African great apes and apart from Asian orangutans. However, after coming to this conclusion, we still had no idea about how to resolve the trichotomy between human, chimpanzee, and gorilla (Fig. I.1). The results from molecular genetic research were confusing, with some studies showing that gorillas were closer relatives to man than chimpanzees, while others suggested the opposite. Based on my observations, I became convinced that chimpanzees were humanity's closest living relatives. This conviction was validated by later research comparing the DNA of humans, chimpanzees, and gorillas, which showed that chimpanzees and bonobos (genus *Pan*) are more closely related to us than are gorillas

Fig. I.1 All African apes perform the 'aeroplane' pattern.

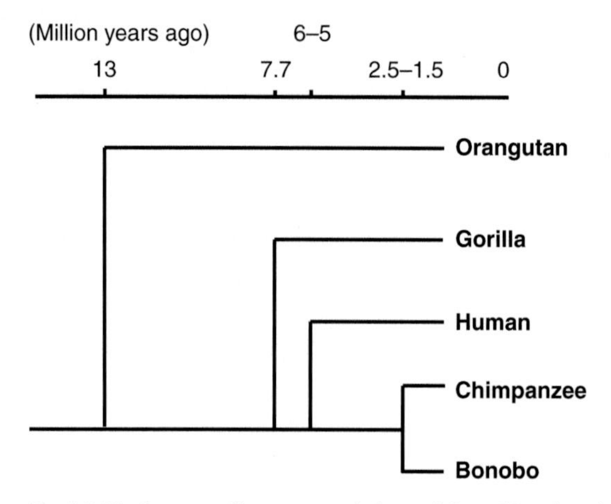

Fig. I.2 Phylogeny of great apes (adapted from Horai *et al.* 1992).

(Fig. I.2) (Horai *et al.* 1992). This success of molecular taxonomy helped to revive my interest in attempting to reconstruct the evolution of human behaviour, a programme of study that I had almost abandoned.

Humans are more similar to chimpanzees than gorillas in many behavioural ways: adult males bond with one another; females rather than males leave their natal groups (patrilocality); both use tools in a variety of contexts; both hunt mammals and eat meat. Furthermore, sex differences in body size in humans and in chimpanzees are relatively small. In sum, behavioural differences support the molecular evidence. Adriaan Kortlandt (1962) had anticipated these findings when he asserted that chimpanzees were human's closest living relatives, long before the evidence from the molecular taxonomy firmly established this fact.

Thus, the results of research on hominoid evolutionary relationships and behaviour began to converge in the mid-1980s. Our ability to reconstruct the behaviour of our ancestors was revived and has prospered. Based on current data, it is likely that the LCA of humans and the genus *Pan* were omnivores, lived in multi-male social units with female transfer, and had antagonistic intergroup relationships. Now I will tell you the story of my exploration into the hinterland of Lake Tanganyika and how I watched chimpanzees.

1

At the beginning

1.1 LAKE AND FOREST

1.1.1 Breath-taking sunsets over Lake Tanganyika

Lake Tanganyika is a long lake that forms a link with the Western Rift Valley of Africa, starting from the Red Sea and ranging from Lake Kivu and Lake Albert to the north, Lake Malawi to the south, finally pouring into the Indian Ocean at Mozambique (Fig. 1.1). With a maximum depth of 1470 m, it is the second deepest lake in the world, after Lake Baikal in Siberia.

Tanganyika is the seventh largest lake in the world, with an area of 32 000 km^2. It hosts both freshwater and saltwater fish. The saltwater fish include four species of sardine called '*dagaa*'. On nights when the moon is new, you can see men rowing along the water holding high-pressure kerosene lamps called *karabai*, scooping up dagaa with massive ladle-like nets. Or in the late afternoon, you can see fishermen in schooners dragging nets across the lake's surface. Trapped dagaa are dried in the sun on the sandy beach and sent to the markets of Kigoma, after being packed in sisal bags.

There are several species of mammal and large reptile, such as otter, hippopotamus, and Nile crocodile, living in the lake. In the reeds by the lakeshore, one can find the African darter roosting. Pied kingfishers put on diving shows, while giant kingfishers, African fish eagles, and bee-eaters perch on nearby boughs.

The lake itself is usually transparent, but occasionally during the rainy season, on the night after a storm subsides, the surface of the lake turns a solid green. If you scoop up some water in a bucket, it is filled with algae and jellyfish. Around March, whirlwinds rise up from the lake. At times, you may see four or five magnificent funnels at once. The sunsets at Lake Tanganyika should be seen in the rainy

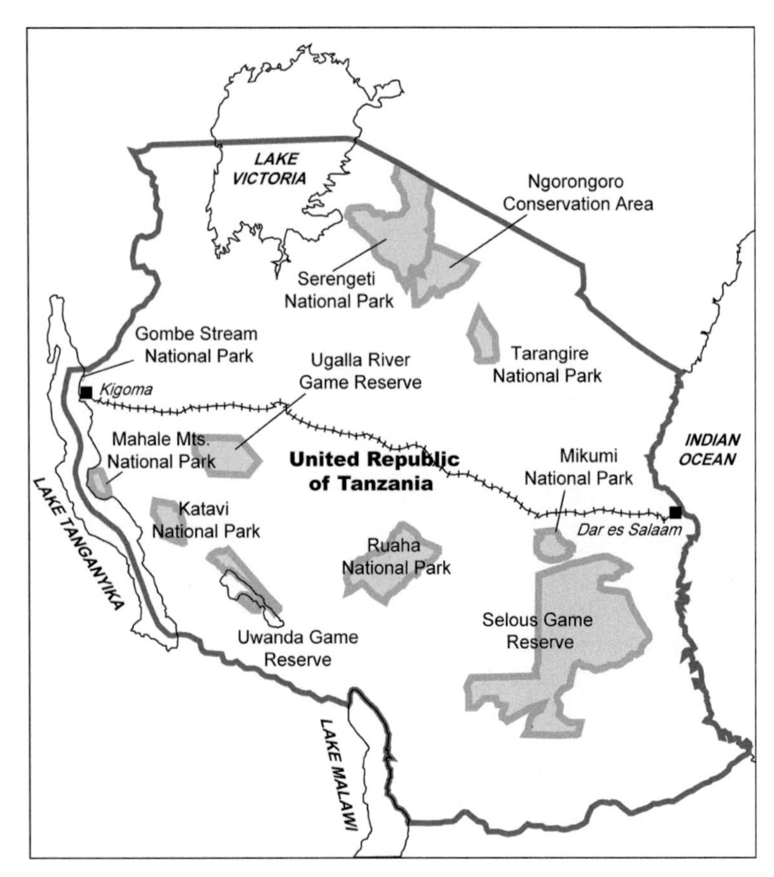

Fig. 1.1 Map of Tanzania.

season; ranging from light orange to crimson violet to scarlet, they are some of the most spectacular in the world.

1.1.2 The chimpanzee forest

The Mahale Mountains are on the eastern shore of Lake Tanganyika, 135 km south of the town of Kigoma (Fig. 1.2). The lake is the region's sole thoroughfare. The nearest roads, which are difficult to travel, are 60–150 km east of the lakeshore, running north from Uvinza and south to Mpanda (Fig. 1.3). From Kigoma, as you move southward along the lake by boat, you see a range of gently sloping hills on the eastern shore. About midway a large peninsula juts out. This is where Kyoto University's first base camp was, at Kabogo Point. A hot spot for sardine fishing, the fishermen's fires dot the shores at night,

Fig. 1.2 Mahale Mountains viewed from the south.

creating an enchanting scene. After the smooth rolling of the hills ceases, you see steep mountains. The northern entrance of the Mahale Mountains National Park is close by.

The Mahale Mountains is one of the world's most spectacular places, with several huge peaks and steep slopes with montane and middle-altitude forests. They are isolated about 200 km from the northern highlands of Burundi-Rwanda and 300 km from the southern highlands of Tanzania. The region is home to several endemic forms of mammals, reptiles, amphibians, insects, and plants.

Western Tanzania is covered mostly with *miombo* woodland (dry forest), which is made up mainly of deciduous trees of the family Leguminosae (Fig. 1.4). Riversides are typically lined with semi-evergreen forest. However, Mahale is over 2500 m at its highest peaks and stretches

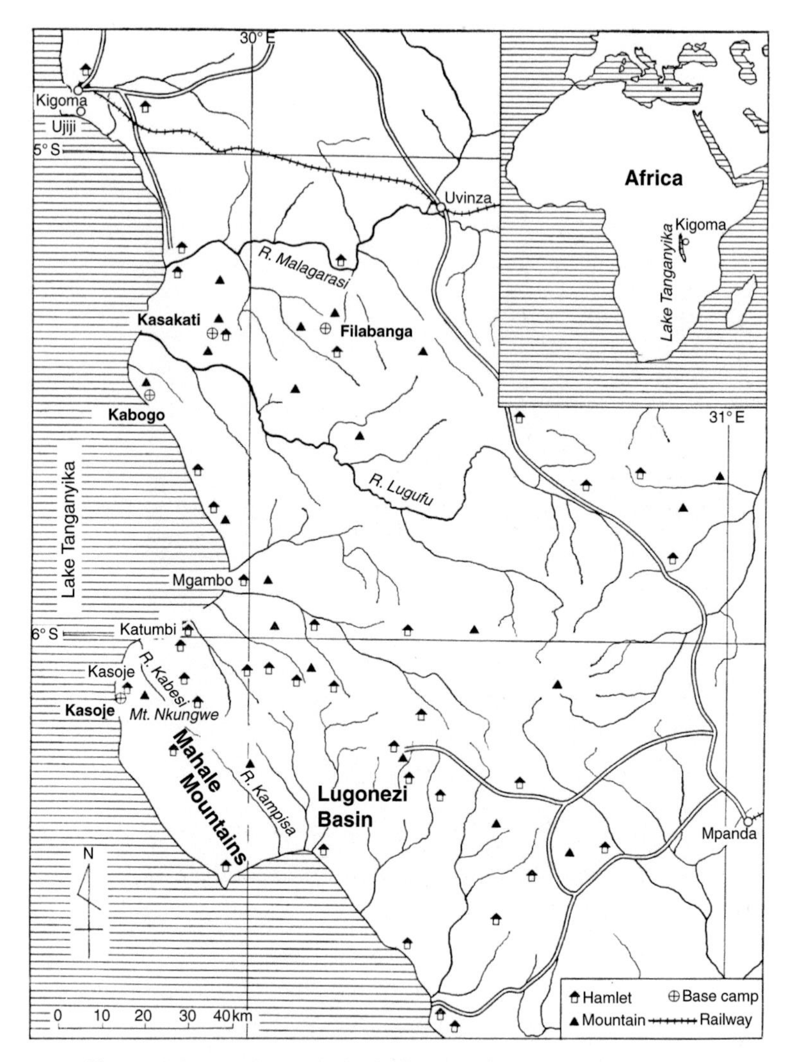

Fig. 1.3 Western Tanzania, including Gombe, Kigoma, Malagarasi, Ugalla and Mahale (adapted from Nishida 1990b).

along the lake from north to south for 80 km. Consequently, a unique vegetation produced by high rainfall and damp air from the lake has developed on the west side of the Mahale Mountains. Annual rainfall ranges from 1700 mm to 2200 mm at the Kansyana camp.[1] I named this tropical, semi-evergreen forest 'Kasoje forest' (Fig. 1.5). Few huge trees

[1] See Nishida & Uehara (1981) for the vegetation, and see Nishida (1990a) for general landscape and climate of Mahale.

Fig. 1.4 Miombo woodland on the slopes of Mt Pasagulu.

Fig. 1.5 Kasoje forest.

exist in the Kasoje forest, but emergents, which do not exceed 30 m in height, include *Canarium schweinfurthii*, *Albizia glaberrima*, and *Parkia filicoidea*.

A large valley pours out into the lake, and this alluvial fan becomes a savanna dotted with acacias and rugged terrain, with rocks scattered here and there. The Mahale highlands are the park's

Fig. 1.6 Mountain forest near the summit of Mt Nkungwe.

Fig. 1.7 Mountain grassland decorated with *Protea gaguedi*.

most beautiful, where mountain forests (Fig. 1.6), alpine bamboo (*Arundinaria*) forest, and mountain grasslands cover the area like a mosaic. The mountain grassland is decorated by beautiful flowers such as *Protea* (Fig. 1.7) and *Helichrysum* (Fig. 1.8).

The area covered by the national park is 1614 km^2 and comprises the Kasoje forest, the Nyenda plateau (Butahya), the Mahale

Fig. 1.8 Mountain grassland decorated with *Helichrysum*.

Fig. 1.9 Solid-stemmed bamboo (*Oxytenanthera abyssinica*).

highlands, Sibuli highland (Bugansa), Masaba-Lufubu Basin, and their surroundings. The last is covered mostly with *Oxytenanthera* bamboo (Figs. 1.9–1.10). Elephant, buffalo, roan antelope, zebra, and other large-hoofed mammals range from Masaba bamboo thicket to Sibuli highlands and use the southern periphery of Mahale, eventually

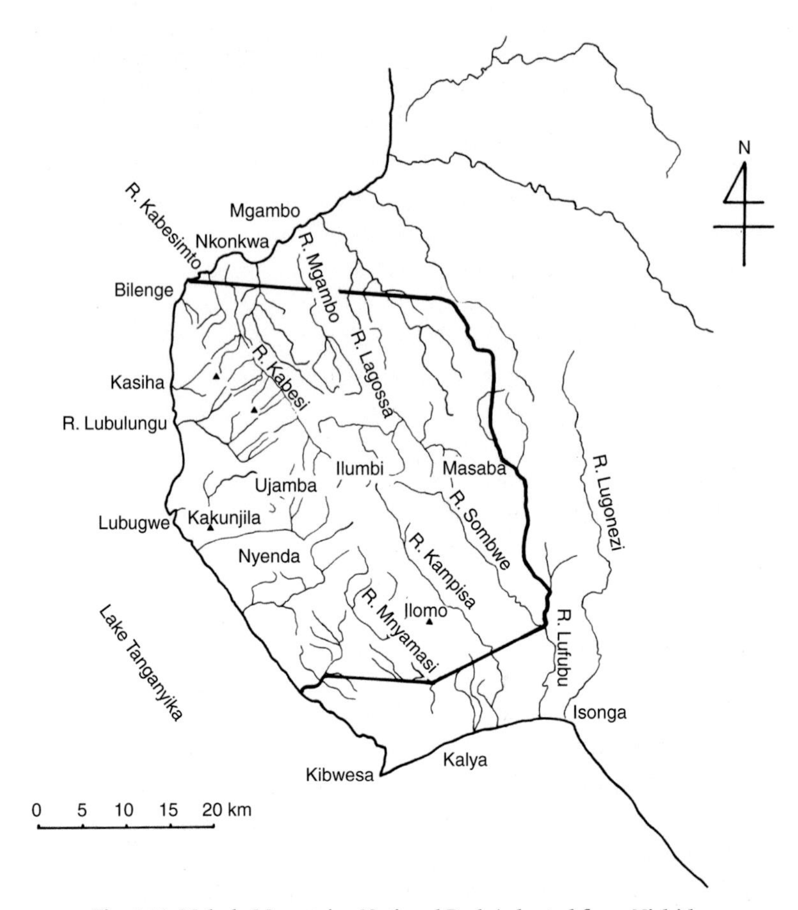

Fig. 1.10 Mahale Mountains National Park (adapted from Nishida 1990b).

ending up at the Nyenda plateau. These large mammals are not found in the Kasoje forest. Instead, Kasoje is home to chimpanzee, red colobus, red-tailed monkey, blue duiker, brush-tailed porcupine, African giant rat, ratel, and crested guinea fowl. The fauna of Kasoje forest resembles that found in forests common to Central or West Africa rather than those of East Africa.

Isolated populations of Angolan colobus and grysbok inhabit the Mahale highlands. Once, when trekking around the eastern side close to the Mahale's main ridge, 2000 m above sea level, I found the corpse of an elephant. This elephant seemed to be trying to cross over Mahale from east to west, when he probably was shot by the mountain-dwelling Tongwe people. His tusks had been taken.

The move to make Mahale a wildlife preserve began in 1974 (Nishida & Nakamura 2008). The next year, Koshi Norikoshi and I were dispatched to Tanzania as experts of the Japan International Cooperation Agency (JICA); we were also assigned as game research officers of the Ministry of Natural Resources and Tourism. In the years to follow, we prohibited the use of fire in the bush, as well as slash-and-burn agriculture. In 1978 the government forced village people to move outside of the proposed national park. Most villagers chose to go to Katumbi to the north or Kalya to the south, the two largest villages of the Mahale Peninsula. In 1979 the Mahale Wildlife Research Centre was established under the supervision of the Serengeti (now Tanzania) Wildlife Research Institute. At that time, patrols of Mahale by its staff were put into full operation.

As one can imagine, the forest in the area began to regenerate. Since game hunting was strictly forbidden, I began to spot wildlife that I had never seen before in Kasoje. First there was warthog, with buffalo, lion, and hunting dog following thereafter.

In 1985 the area of 1614 km^2, including the Mahale Mountains, was designated as a national park. This was 16 years after I had first made a proposal to Mr J. S. Capon, the acting director of the Game Division. At the time, we estimated about 700 chimpanzees lived in the park. In the 1960s Takayoshi Kano and I surveyed most of the park's area. We recorded where we heard chimpanzees call, their beds ('nests' is a misnomer, see below), and also sightings of chimpanzees. From this we determined that there were about 14 unit-groups or communities in Mahale. We arrived at our estimate of 700 chimpanzees (Nishida 1968; Kano 1972) by assuming that about 50 chimpanzees lived in each group.

1.2 THE AFRICAN APE RESEARCH PARTY SETS OUT

1.2.1 The Japanese are going to build an airfield?

Kinji Imanishi organised the first Kyoto University African ape expedition in 1961. As a member of the fourth expedition party, I began research at Kasoje at the northern foot of the Mahale Mountains in October 1965. The previous year, the Tokyo Olympics were held and the *Shinkansen* (bullet train) began running between Tokyo and Osaka. In 1967 the Japanese embassy in Tanzania was established at Dar es Salaam, while simultaneously the first members of the JICA (40 young women) were deployed to teach needlework in various parts of the country.

High economic growth in Japan between 1955 and 1965 made it possible to fund overseas scientific research and development aid in places such as Africa for the first time since the end of the Second World War. Research on chimpanzees could not have begun at a better time. Kyoto University's African ape expedition conducted research at Kabogo base camp from 1961 to 1963.

After Imanishi and Itani decided to change their target of study from gorillas to chimpanzees in 1960, they became interested in how our human ancestors adapted to a savanna habitat. This change was influenced partly by the prevailing idea in anthropological thought of the savanna as the cradle of human evolution. With this change in emphasis, the research team incorporated ecological study with Imanishi's sociological research into the origins of the human family.

On the way back to Kyoto from Africa in 1960, Itani consulted with Louis Leakey about possible research sites to study chimpanzees. Leakey proposed Gombe to him and explained: 'A young British girl, Jane Goodall, is now studying chimpanzees there, but she will stop her research in six months and return to her country. Then you can enter the area.' Itani had already called on Goodall at Gombe and knew that it was located in an area including savanna. He decided to follow-up on Leakey's suggestion.

Thus, the research team's initial destination in 1961 was not Kabogo, but Gombe. However, when Itani called on Leakey in Nairobi in early November 1961, the latter informed him that Jane Goodall had decided to continue her research and that Imanishi's team had to look for a new field site (Katayose 1963). Twenty tons of equipment, books, and Japanese food, including three tons of cement (as they did not think cement could be obtained in Tanzania!) from Japan had arrived at Kigoma. The research team had to pay a considerable sum of money to store these goods every day. A game officer of Kigoma recommended several sites including Kabogo and Mahale, and when Imanishi and his crew surveyed Kabogo point, he was satisfied with the site, as it had savanna with a developed riverside forest and some elevated hills. The latter was particularly attractive to Imanishi as he had not forgotten that he was an alpinist!

Shigeru Azuma and Akisato Toyoshima (now Nishimura) began research at Kabogo. They tried to provision the chimpanzees by feeding them bananas and honey. They found that the chimpanzees fissioned and fused into subgroups, but there were few opportunities to observe them. As a result, they could not clarify the social unit of chimpanzees (Azuma & Toyoshima 1962). Given scant progress in the habituation of chimpanzees at Kabogo, Itani moved the base camp to the Kasakati Basin in the latter half of 1963 (Fig. 1.3). In 1965 my classmates, Kohsei Izawa and Takayoshi Kano, and I went three separate ways and conducted research at Kasakati, Filabanga, and Mahale (see Section 1.2.3). All of these camps were of African style, with mud walls and grass-thatched roofs. They contrasted sharply with the modern base camp full of equipment at Kabogo. Ultimately, only Mahale remained as a long-term research site. Only there was it possible to habituate chimpanzees to human presence and observation (Nishida 1990a).

Chimpanzees had been known to the scientific world to exist in Mahale since 1935. Articles about their distribution around Mahale by Dollman, Hatchell, Moreau, McConell, and Grant appeared between 1935 and 1946. Kano pointed out that Moreau's 1942 description proved to be surprisingly correct, even though Moreau himself had never visited Mahale! Mahale was not the first candidate for a chimpanzee project, because it was many hours' voyage from Kigoma, the nearest point of contact with civilisation.

The chimpanzees of Tanzania live at the southeastern extent of the species' distribution. To the north of Mahale, Gombe chimpanzees are connected with the Ugandan population through those of Burundi and Rwanda. On the other hand, Mahale chimpanzees seem to be connected with the populations to the south of the Malagarasi River (i.e. the upper valleys of the ancient Congo River) through Karobwa, Kasakati, and Ugalla. In 1996 Hideshi Ogawa and his colleagues (1997) discovered a new population of chimpanzees in the Lwazi River, far to the south of Mahale.

I regret that I was unable to see Kabogo base camp while it was still active. When I visited it for the first time in 1967, there were only two prefabricated building structures left standing. At one time, there had been two Toyota four-wheel vehicles, two generators, four refrigerators, a gas cooking stove with twenty-five 100 kg canisters of propane gas, three microscopes, five ladders, and an immense library. Imanishi had planned to do long-term research and set out to build a permanent base camp (Fig. 1.11). A builder had even accompanied him all the way from Japan, and many of the local people had become involved in the construction of the camp. At the time, a rumour spread that the '*Wajanpani* (Japanese) are going to build an airfield!'

Fig. 1.11 Professor Kinji Imanishi at Kabogo camp (courtesy of Junichiro Itani).

1.2.2 From Kyoto to Kasakati and to Mahale

I left Osaka Airport for Dar es Salaam, capital of Tanzania, with Junichiro Itani and Takayoshi Kano on 7 July 1965. We arrived there via Bangkok, Karachi, and Nairobi on 15 July. We had to spend three weeks waiting for a huge cargo shipment from Japan. Unfortunately, there was a strike by dock workers, and it was impossible to retrieve the cargo from the port at Dar es Salaam. We also spent much time negotiating to buy a rifle. The expedition already had two rifles, but Itani, the expedition leader, considered it necessary to purchase another. There were several regulations about foreign persons owning a rifle, and few Tanzanian officials knew how to deal with this. We had to move from one office to another to find the responsible officer. I found most of the stay in Dar es Salaam unpleasant. Due to this experience, I vowed never again to buy a gun or to bring huge supplies from Japan.

After I spent about a week buying food and camping supplies in Kigoma, on 16 August I arrived in the Kasakati Basin, where my colleagues, Kohsei Izawa and Akira Suzuki, had been studying chimpanzees. In the initial plan, three students, Izawa, Kano, and I, were to establish different camps in the Kasakati Basin and to follow the same chimpanzees. However, Itani changed this plan, instead deciding to send us to three different sites: Kasakati, Filabanga, and Kasoje (Mahale). There were two reasons for this change of plan. First,

Fig. 1.12 My 'father', Saidi Sobongo.

provisioning at Kasakati had been unsuccessful because elephants had damaged banana plantations planted for chimpanzees. Second, the population of chimpanzees was estimated to be too small for three students to study.

On 2 October I left Kasakati for Kigoma, and after shopping for a few days, I took the steamship *Liemba* for Mgambo (Lagossa in Fig. 1.10). We left in the evening, and the journey took 12 hours. I was still 30 km from Kasiha (a hamlet of Kasoje), my final destination. However, I had neither a boat nor an outboard engine. An Arab merchant, Said Seif, with whom I had shared the same first-class room on the *Liemba*, provided me with lodging in his house for three days and lent me a small boat with a 3 hp 'seagull' engine.

At 22:30 hours on 11 October 1965 I arrived at the shore of Kasiha village (Fig. 1.10). More than three months had passed since my departure from Japan. When I awoke early the next morning, I found before me an extremely beautiful mountain range, richly coloured with forests and montane savannas. I called on and greeted Mr Saidi Sobongo. Kohsei Izawa had kindly arranged for him to help me when he made a preliminary survey to Kasoje in early September. Saidi was a middle-aged son of the village headman (*mwami*), Sobongo (Fig. 1.12). For the first few days I stayed in a village 'guest house': one room with a kitchen, but without a door! Saidi later invited me to move into his house. His house consisted of three rooms under an oil palm grove. The left room was occupied by one of his wives, Binti Sudi, while another wife, Wantendele, lived in a room on the right side of the house. Saidi himself was supposed to occupy a middle room between both wives, but he did not use it, as instead he slept with one or the other wife on a one-week rotating basis.

On the day I arrived, I asked Saidi to show me a possible plot of land where I could plant sugarcane for the chimpanzees. Ideally, the spot should be frequented by the apes. Saidi took me to Kansyana Valley, about 1 km east and inland from Kasiha village. There he showed me a large area of elephant grass that previously had been planted with maize but had been deserted a few years before. People did not use the same plot of land for more than a few years, and they did not reuse it until at least three decades had passed. As slash-and-burn farmers, the only fertiliser utilised was the ashes of trees cut at the time of the forest clearance. There were many chimpanzee trails and wadges of the pith of the elephant grass. I heard chimpanzees call almost every day from the village. Even on the first day of my arrival, I approached a group of chimpanzees and observed them in the bush for 17 minutes. The plot of land was surrounded with old secondary forests on both the western and eastern sides. Thus, I surmised that chimpanzees who appeared in the plantation could easily find places to hide if they were anxious or scared. I felt that this was an ideal place to attempt to provision and habituate wild chimpanzees.

The rainy season was approaching, and as soon as possible before the rains came we had to clear and burn the bush to plant sugarcane. I hired 40 people to clear the land, cut trails, cook food, and track the chimpanzees. In the beginning, sugarcane planting was surprisingly easy! Just buy a big (mature) cane and bury half of it in the soil. That's all! New shoots germinate from buds of each joint and it grows without any fertiliser. As they grew, we had to plough nearby and add soil to saplings.

When sugarcane harvesting time approached, termites and caterpillars began to attack the plants. I dealt with these insects by removing them by hand. Cane rats also began to eat the sugarcane, but their damage was minimised by removal of weeds, elephant grass in particular. Cane rats avoid even partially open areas as they are extremely afraid of leopards. There was a severe problem in the form of nocturnal raiding by bushpigs. Bushpigs consumed 60 sugarcanes in one night! People advised me to shoot them, set wire traps, enclose the whole field with a palisade stockade, or to employ a night watchman. I deemed all of the measures except the last one inappropriate. Even if I obtained a gun, the sound of shooting would have alarmed the chimpanzees. Wire traps would also be dangerous to them. Fence-making would have required a lot of time and money and may have inhibited chimpanzees entering the plantation. In the end, I asked the uncle of Saidi to work as a watchman. However, crops continued to be raided every night, as he slept soundly through the raids by the bushpigs! As a consequence, I spent several weeks alone in a small hut in the plantation before I found an appropriate guard.

I started at Kasoje with virtually nothing. A small tent, canvas safari bed, sleeping bag, water bottle, hango or Japanese rice-cooking pot, simple cooking set, medicines, 70 field notebooks, pair of binoculars, Nikon camera with 300 mm telephoto lens, mini tape recorder, 100 rolls of film, 20 m of nylon cord, 100 small nylon bags, 10 boxes of mosquito coils, 3 pairs of caravan shoes, soy sauce, and tinned food bought in Kigoma. That was it.

In addition to the plantation, I put some mature sugarcane along and close by some of the newly cut observation paths. Chimpanzees began to eat this sugarcane in the middle of December 1965, although I never saw them doing so. Finally, in March 1966, some chimpanzees began to raid the sugarcane plantation. At that time I was living at Nganja, attempting to follow chimpanzees 7 km south of Kansyana camp. I returned quickly to Kansyana to confirm how provisioning had been proceeding.

On 27 April 1966, after spending six months living in Saidi's house, I moved to Kansyana, 1 km inland from Kasiha village. My new dwelling was a straw hut that was built in a day (Fig. 1.13). In it was a log bed (that took up half of the hut), and a tiny wooden table a local craftsman had made for me. When the table was placed in the hut, there was very little room to move, or even breathe!

On 13 May I saw a black-haired arm breaking sugarcane in the plantation for the first time. I began to record physical characteristics of individuals: their sex, body size, hair colour, baldness, wounds, presence or absence of white beards, and other traits. Then I gave them names. I borrowed Sitongwe names for the chimpanzees so my Tongwe assistants

Fig. 1.13 Kansyana camp in 1966.

could remember them easily. Two daughters of Bunengwa rushed to my camp. They protested when they heard that I had given one of the adult female chimpanzees the name of *Gwabunengwa* (meaning 'daughter of Bunengwa'). In spite of this, I named two juvenile female chimpanzees after their own maiden names, *Tatu* and *Ndilo*.

Let me explain my naming principle. First, I took names only from Tongwe and Swahili so that my assistants might learn easily, but I also began to use Japanese, as some locals seemed to suspect that I considered chimpanzees to be more like Tongwe people than Japanese people! As it turned out, my assistant named a silly young adult male *Nishida* when I was in Japan. As chimpanzees continued to be identified, I limited the number of letters to seven and gave each chimpanzee a two-letter abbreviation. I also established a policy of naming infants only after they had reached two years of age, because young infants suffered high rates of mortality. (I recently learned that Cynthia Moss (1989) had a similar policy of naming elephant calves only after they reached four years of age.)

At the time my budget was 1200 shillings per month, which is about 60 000 yen or 170 USD. This was supposed to cover the salaries of seven African staff, gasoline expenses, sugarcane for provisioning the chimpanzees, plus my food. I got on frugally. After I spent about six months at Kasoje, I lost my wristwatch. They sold watches in the township of Kigoma, but they were 100 shillings (equivalent to 15 USD) each, so I could not purchase one. A few months later, Itani came over from Japan. He was shocked to discover that I was 'watchless' and kindly

Fig. 1.14 A Tongwe farmer, in the woodland of the Nyenda Plateau.

bought me a new one. But as luck would have it, I lost it a few days later. I did not think it prudent to ask him for another watch, so I spent the next few months without one. Because of this, the observations I made in my doctoral thesis lacked times and did not have a record of total observation hours. What was important was to record social behaviour and interactions between chimpanzees. Luckily, my thesis review committee failed to notice that my thesis had no records of time! The committee members seemed satisfied that I clarified the basic social structure in wild chimpanzees.

Despite my spartan living conditions, I did have one spectacular thing going for me: sharing the community with the villagers of the Tongwe people. Tongweland in those years was like a living museum. The men of the village were naked from the waist up. When they left the village, they always walked spear in hand (Fig. 1.14). When I left the village alone with only a *panga* (large bushknife) some adult males praised me, saying 'You are truly a brave man going to the bush without a spear!' The possessions of families around the village were varied: unglazed water jugs or pots, stone mills and mortars, wooden mortars and pestles, woven mats made of date leaves, grass, or bamboo, bamboo colanders, small stools called *sitebe*, and bellows made of wood and colobus fur. Everything was hand-crafted, of course.

There was little in the way of commerce. Everyone walked around barefoot, and as far as clothing went, the women wore *kanga* (two pieces of rectangular cloth worn on the upper and lower part of the body). The children were naked, and the men wore clothing that barely passed for rags. This appeared pitifully unreasonable, because they still had their traditional technique of making bark cloth from fig trees, but village women hated to wear these dull-coloured clothes and just threw them away after seeing colourful *kangas* worn by town women. Many of the bare necessities that one takes for granted (e.g. sugar and soap) were usually lacking in households of the Tongwe; some households even lacked salt. Everyone had hand-made gill nets for catching small fish in the river, and a few people had trap-like devices for catching giant catfish, called *nsinga*, in the deep lake, but not a single person could afford to buy these things made from commercial fibre. What people were most keen on in my belongings was a metal kerosene lighter made in Austria, because they always had to keep a fire lit, as they did not have matches.

I did not find fault with the people's lives. In fact, I felt the opposite. The self-sustenance of the Tongwe left me awestruck; they could make practically anything by hand. They were jacks-of-all-trades, and excelled at many tasks. As long as you were not too particular about quality or design, they could put together anything for you, from a house to a henhouse. They did not have nails or metal hammers – these items were available at the shops of Arab merchants 20 km away but were too expensive. Instead, they collected several species of woody vines from the bush and used them, for example, to connect wooden pillars with bamboo crosspieces.

The town of Kigoma was the nearest point of the rest of the civilised world to Kasoje. Twice a month the steamship *Liemba*, would travel past Mahale between Kigoma and Mupulung, a port to the south, in Zambia. To reach the landing where the steamship made landfall you had to take a six-hour sailboat ride from Kasoje to Mgambo. Airmail from Japan took at least a month to arrive, so one can understand the isolation I suffered, living in such a foreign land.

The first two years or so living in Kasoje left the most long-lasting impressions on me. If not for the cheerful and open attitudes of the Tongwe people, there is no way I would have been able to survive and do my research. They would amuse themselves by dancing into the wee hours of the night, and if the drinks were on you, they would sing and dance for you anytime you liked. Those first two years at Kasoje affected the way I saw things forever.

1.2.3 At the feeding place

Before chimpanzees began to raid the sugarcane plantation, we had begun to put out 10–40 pieces of sugarcane at the fixed feeding place 200 m upstream from my camp, along the Kansyana valley. My observations were mostly conducted from a scaffold 2 m high, built 40 m from there. I continued to buy sugarcane before and after establishing the plantation, in order to facilitate provisioning. Now, all the sugarcane produced at the village of Katumbi was exhausted, and we had to send my team to villages farther and farther away. In the end I sent people even to Kalya (Fig. 1.10), 80 km south of Kasiha! I was devastated when Omari Kabule did not bring any sugarcane from Kalya because the price was too high. If he had taken the cost of gasoline and labour into consideration, he would have bought the sugarcane, but his only concern was to obtain the item for the customary price.

Finally, we signed a contract with a rich farmer in upper Mgambo, who agreed to sell us all the sugarcane he could produce. However, we found it difficult to continue this system for very long, because the amount of sugarcane we needed varied from month to month. I then thought about changing the food used to provision the chimpanzees from sugarcane to bananas. A banana plantation would be much easier to maintain. After planting saplings, all that would be required would be removal of the occasional weeds with hoes. Bushpigs, cane rats, and insects do not raid banana plants. Moreover, bananas are much easier to transport than sugarcane, which is heavy. I was able to carry only five mature canes from the lakeshore to the Kansyana camp in one trip. Banana trees grow continually from clones in the ground after the parental tree is cut. As it turned out, unripe bananas were never eaten by chimpanzees, although piths of stems were occasionally eaten by hungry chimpanzees.

On 31 July 1966, I hung a bunch of very ripe bananas on a shrub, in addition to sugarcane scattered on the ground at the feeding place. The first chimpanzee to appear that day was an old female, *Wankungwe*, who came alone, stopped before the bananas, gazed at them for several seconds, and then began to eat. I was surprised because I had been told that village people had never seen chimpanzees eating banana fruits in their banana groves. After *Wankungwe* consumed all the bananas and disappeared, I hung the next ripe bunch and waited. After four hours, four adult males and *Wankungwe* appeared at the feeding place. A young adult male, *Kaguba*, passed by the bananas, as if he did not notice. However, the second individual, *Wankungwe*, and three other adult males stopped in front of the bananas and began to jostle with one another, all of them gazing at the fruit. Finally, *Kamemanfu* sat before the bananas, with *Wankungwe* staying nearby. In turn, *Kajabala* and *Kasanga* almost covered *Wankungwe*'s back. All were panting violently. *Wankungwe* stepped in first and began to eat the bananas. Then *Kamemanfu* and the others followed suit, as if competing with one another. This must have been the first time they ate bananas in their lives, because bananas are harvested by humans long before they become ripe. *Wankungwe* seemed to realise that the fruits were edible from their fragrant smell. Adult males might have eaten eventually without watching *Wankungwe* do so – however, it is obvious that seeing her eating led them to eat sooner. Following this episode, we began to provide sugarcane and bananas at the feeding station.

1.2.4 Discovery of social units

Varied groups of chimpanzees appeared at the feeding place. One day, only three adult males came together in the morning, with two mothers and their offspring appearing in the evening. The next morning, these males and mothers came together. One of the adult males and one of the mothers came later in the evening. The composition of chimpanzee parties was extremely variable, and I continued to record the physical (particularly, facial) characteristics of each individual. After about three months, I noticed that the number of identified chimpanzees had reached 21. The count did not increase until the next year. Subgroups seemed to occur just among these 21 chimpanzees. I hypothesised that these 21 individuals formed a social unit or group; I named the group K-group after one of the most habituated adult males, *Kajabala*.

In September 1966, K-group's chimpanzees stopped visiting the camp so often. The whole group moved northward on 4 November. A larger party came from the south and began to use the area, including the feeding place. At first I thought new members of K-group had finally come to Kansyana and that they would mingle with them. I searched for K-group chimpanzees, but I saw none of them in this new party. None of the newly arrived chimpanzees were tame at all, and I realised that there were two groups or social units using the same feeding place. I began to suspect that the chimpanzees who had just arrived were those I had been following in the south the year before. I had already identified several adult males the year before; one of them was *Mimikire* ('ear cut' in Japanese), so I named this bigger group 'M-group'. I returned to the naturalistic method of simply following chimpanzees in the forest. K-group seemed to have left our camp area permanently.

I returned to Japan at the end of March 1967, after a 19-month stay in Tanzania. I remained in Japan for three months. I married Haruko Kitayama, who moved from her hometown, Utsunomiya, to Kyoto to live with my parents in their house.

On 1 July 1967 I left Japan, again heading for Kasoje. I arrived at Kansyana on 14 July. Ramadhani Nyundo, who had been feeding chimpanzees, told me that K-group had begun to visit Kansyana in April. I confirmed the identities of all of the 21 chimpanzees I had identified the year before. From this, I determined that the multi-male, multi-female group seemed to be stable in composition for at least a year. In early September 1967, K-group disappeared again, as soon as a big party of M-group came towards our camp from the south. This time I left Kansyana to search for K-group to the north of the area it used from April to September. I was able to find the chimpanzees of K-group in the Myako valley, 3 km north of Kansyana. Taking advantage of the fruiting season of *Garcinia huillensis* that occurs only locally, I, along with my assistants, made surveys of other chimpanzee groups. We found about six unit-groups in the Kasoje area (Fig. 1.15).

I was able to learn that the chimpanzee social unit comprised many adult males, many adult females, and immature individuals. Adult females appeared to outnumber adult males 3:1. Individuals in this social unit fissioned and fused, but moved over a fixed territory. Members of different groups had antagonistic relationships, with M-group dominant to K-group. After spending four months at Kasoje in 1967, I returned to Japan. It was time to write my PhD thesis.

1.2.5 M-group provisioned

On the evening of 4 August 1968, I arrived at Kasoje for the third time. The purpose of this visit was to investigate the relationships between K- and M-group.

The next evening I was surprised to hear K-group immediately respond with pant-hoots to Ramadhani's call as he simulated the apes' calls, 'Uh-ho, Uh-ho.' While I was in Japan, he had brought sugarcane to the feeding place and 'pant-hooted' every morning. K-group's chimpanzees had learned that Ramadhani's calls meant the arrival of sweet food at the feeding place.

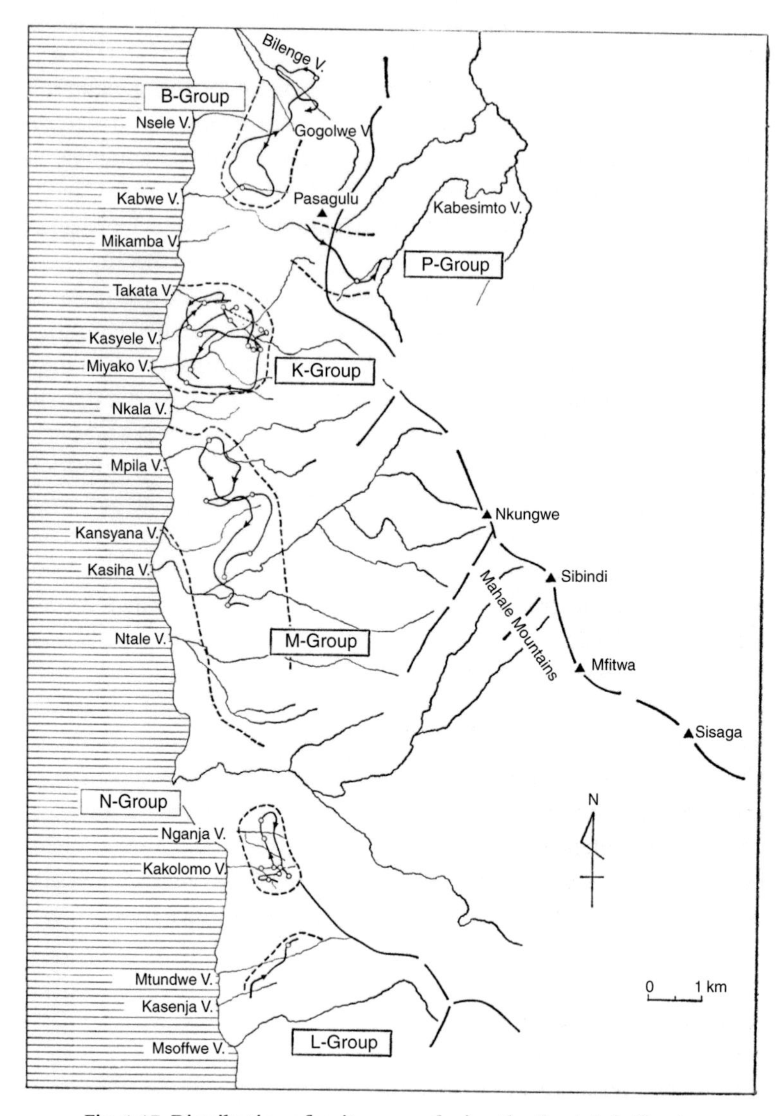

Fig. 1.15 Distribution of unit-groups during the *Garcinia huillensis* season of 1967 (adapted from Nishida 1973a).

This gave us a convenient method of tracking chimpanzees. A tracker and I carried sugarcane pieces or bananas in our backpacks, and from a high vantage point, we pant-hooted and chimpanzees responded to us with an outburst of calls. We then approached them, and they approached us in cooperation! When we met chimpanzees, we scattered sugarcane pieces or bananas and checked the composition of the party. After they consumed the food, we simply followed them. I used this 'mobile provisioning' method for K-group chimpanzees until 1982.

Some chimpanzees of M-group gradually began to eat sugarcane at the same feeding place at Kansyana, and we successfully provisioned individuals in M-group by early November 1968. However, the process of full habituation of M-group took many years. It was not until 1981 that Mariko and Toshikazu Hasegawa finally habituated and identified the last female and her family.

During the 1968 field season I found two adolescent females, *Chausiku* and *Tausi*, and the old female, *Wankungwe*, of K-group mingling and travelling together with M-group chimpanzees. I saw no antagonistic responses to the newcomers by resident M-group chimpanzees. Instead, the K-group females were behaving as if they were M-group members! *Wankungwe* often appeared at the feeding place alone, so her identity was rather ambiguous. However, the two adolescent females continued to follow M-group parties and never seemed to return to K-group. Thus, I speculated that the adolescent females might be transferring from K-group to M-group, and this might be the basic difference between hominoid and cercopithecoid society. The reason I adopted this hypothesis so early was that I had studied the emigration and immigration of adolescent males among Japanese macaques for my master's degree before I came to Africa.

It was the first case of female transfer recorded in non-human primates, and after a few years, we established that chimpanzees have a patrilocal residence system. This differs from the matrilocal system in monkeys, such as macaques and baboons. Later, a patrilocal residence system was also described in New World monkeys, such as spider monkeys.

Although my research lagged five years behind Jane Goodall's study at Gombe, the reason I succeeded in unravelling the social unit of chimpanzees was partly fortuitous. The feeding place was accidentally sited in the overlapping zone of the territories of two unit-groups. Thus, it was easy to observe a subordinate group supplanted by a dominant one. Moreover, earlier research in the Kasakati Basin and Filabanga had already suggested that chimpanzees lived in large-sized groups.

When I published part of my doctoral thesis in 1968, Western primatologists had different views. Goodall stated that 'the only group that may be stable over a period of several years is a mother with her infant and young juvenile offspring' (Goodall 1968, p. 167). Reynolds (1966) published similar findings, suggesting that chimpanzees lived in 'open groups' that lacked a defined social unit. The assertions made by

Western primatologists were generally accepted at the time. My discoveries were neglected by most researchers in the field for at least five years. Goodall (1973) finally acknowledged a large group as the social unit of chimpanzees in 1973. The rest of the academic world followed only in 1979.

In the chapters that follow, I relinquish the journal-style description adopted here. It is impossible for me to recount in detail occurrences in chronological order, as my field study has unfolded over more than 40 years! Instead, I will note behavioural differences among study sites and emphasise Mahale chimpanzee culture whenever appropriate.

2

Food and feeding behaviour

2.1 FISSION AND FUSION OF PARTIES RANGING FOR FOOD

2.1.1 'Congregating' season versus 'dispersing' season

The subjects of our long-term research were K- and M-groups, whose home ranges extended through the northwestern foothills of the Mahale Mountains, occupying areas of 10 km^2 and 30 km^2, respectively, in the middle of the Kasoje forest (Fig. 2.1). The numbers in K- and M-groups in the 1970s were about 30 and 80 individuals. Table 2.1 shows a breakdown of the groups' composition (Hiraiwa-Hasegawa *et al.* 1984). The size of M-group increased to about 100 after most of K-group's females immigrated to M-group in 1979 (see Chapter 6).

Unlike many other kinds of non-human primates, chimpanzees from the same group do not always travel together. They break up into several smaller groups when foraging, while on other occasions they meet up, interact, and form a larger group. For example, in September and October 2002, all of M-group (54 individuals at that time) travelled together for 17 of 57 (30 per cent) observation days. This congregating differs from what has been said about the unit-groups (communities) at other sites, where the entire group never gathers together (Goodall 1986). However, a subgroup typically does not remain stable in membership, and at any given time it can have almost any composition (Nishida 1968). The size of such a 'party', as these subgroups are called, depends upon several factors: amount and distribution pattern of staple foods, size of food patches, absence or presence of oestrous females, absence or presence of conflict among high-ranking males, etc. (Sakura 1994; Matsumoto-Oda *et al.*1998; Hosaka & Nishida 2002; Lehmann & Boesch 2004; Turner 2006). Moreover, Lehmann and Boesch found that

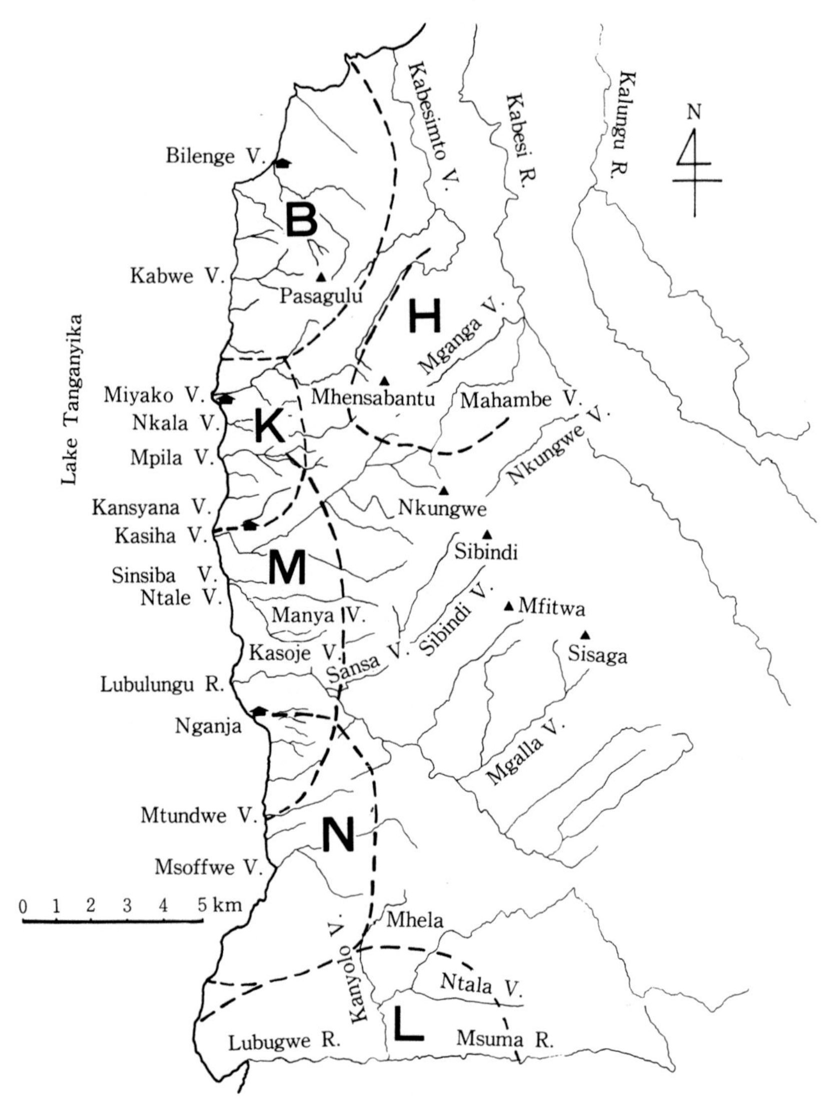

Fig. 2.1 Map of B-, K-, M-, N-, H- and L-group ranges (from Nishida *et al.* 1990).

Table 2.1 *Age–sex composition of K- and M-groups at Mahale*

Group	Year	Adult		Adolescent		Juvenile		Infant			Total
		Male	Female	Male	Female	Male	Female	Male	Female	Sex unknown	
K	1967	6	9	1	3	0	1	3	6	0	29
	1972	5	13	0	6	1	4	4	1	1	35
	1977	3	14	1	4	2	1	5	2	0	32
	1982	1	5	2	0	1	0	0	2	0	11
M	1977*	10	12	5	1	0	0	2	3	0	33
	1982	11	39	9	10	3	10	9	12	2	105
	1987	8	33	9	9	6	11	3	5	0	84
	1992	9	30	7	6	4	7	7	5	1	76
	1997	7	18	4	3	3	3	2	2	2	44
	2002	7	18	5	2	2	3	3	9	0	49
	2007	10	18	3	7	3	6	0	5	3	55

* Identifications of M-group's chimpanzees were not complete

party size, party duration, and male–female association increased as study group size decreased in Taï.

Although the exact periods vary depending on the year, a typical year can be roughly divided into the 'congregating season', when the group often moves as a single large unit, and the 'dispersing season', when the group breaks up into many smaller parties. At Mahale, the gathering season typically begins in June (early dry season) and finishes in the middle of January (mid-wet season) the following year. In other words, the congregating season accounts for 4.5 months of the dry season and 3 months of the rainy season. The dispersing season is thus the remaining months, the latter part of the rainy season.

2.1.2 Menu of chimpanzees

In mammals that are closely related, 'the bigger the mammal, the poorer the food' is a rule of thumb. Since larger-bodied mammals have a smaller body surface area relative to their body weight, their heat loss is relatively smaller. Therefore, even by consuming foods as low in calories as shoots, leaves, stems, and bark, large-sized mammals can still survive, while small animals cannot.

Apes mostly eat fruit. We know that fruit is rich in sugar and has a high nutritional value. If we compare diets among apes like gibbons, bonobos, chimpanzees, orangutans, western gorillas, and mountain gorillas, we see that as body size gets larger, fruit intake as a component of the diet declines proportionately. The ability of the mountain gorilla to live in high altitudes where fruit is scarce is due to its body size. The staples of the mountain gorilla's diet are bamboo shoots, celery, and leaves from vines, but when it can get its mouth on a juicy fruit, that bulky body will shimmy up a tree for it.

Chimpanzees' chief nourishment comes from fruit, but they also feed on seeds, leaves, buds, shoots, flowers, the pith of stems, inner bark, xylem (dry woody portions), roots, gums, etc. – that is, pretty much every part of a plant. Besides plant food, chimpanzees eat many insects, including ants, termites, bees and their honey, and the larvae of some beetles and flies, as well as vertebrates such as monkeys, galagos, bovids, suids, hyraxes, civets, game birds, and bird eggs (Nishida & Uehara 1983) (see Section 2.4). Finally, the apes take one or two mouthfuls of termite soil on a regular basis, perhaps several times per month. They also lick rocks in larger valleys and along Lake

Tanganyika, particularly in the dry season. They may lick for more than an hour, but the mineral that they ingest is not known. Chimpanzees are omnivorous fruit-eaters.

Clearly, the most important constituent of the chimpanzee diet is fruit. More than 80 species of fruits are eaten by M- and K-group chimpanzees. The most important species are listed here. April–July: *Cordia millenii, Afrosersalisia cerasifera, Monanthotaxis poggei, Psychotria peduncularis, Uvaria angolensis, Arbotrys monteirode*, and *Harungana madagascariensis*; August: *Cordia africana, Ficus vallis-choudae*, and *Voacanga lutescens*; August–October: *Uapaca nitida, Syzigium guineens, Pycnantus angolensis, Pseudospondias microcarpa, Garcinia huillensis*, and *Saba comorensis*; October: *Parkia filicoidea* and *Saba comorensis*; November–January: *Saba comorensis*.

The fruits of the species mentioned above are abundant during the months of the congregating season, although the actual staple food changes year by year (Nishida & Uehara 1983; Nishida 1991b). Moreover, figs of *Ficus capensis* and *F. urceolaris* are available throughout the year. On the other hand, during the dispersing season, we hardly ever catch a glimpse of any fruit with a high or prevalent yield. Fig. 2.2 shows the results of faecal analyses done in 1975–1976.

Besides fruits, young leaves are important constituents of the chimpanzee diet. Leaves from more than 90 species of plants are eaten, in particular, leaves of tall trees such as *Pterocarpus tinctorius, Ficus exasperata*, and *Sterculia tragacantha*, leaves and blossoms of woody vines such as *Baphia capparidifolia*, leaves of shrubs such as *Ficus urceolaris*, and petioles of herbaceous vines such as *Ipomoea rubens*. In August and September, chimpanzees spend much time (often two hours at a time) eating galled leaves of *Milicia excelsa*. Moreover, herbaceous piths of elephant grass (*Pennisetum purpureum*) and African ginger (*Aframomum* spp.) and woody piths of *Landolphia owariensis* are eaten in large amounts.

According to my long-term records, chimpanzees rarely eat tubers or subterranean organs of plants; in other words, they usually do not eat potatoes and will not give mushrooms even a glance. The only exception is the root of a short tree of the bean family (*Aeschynomene* sp.), growing on the sandy beach of Lake Tanganyika, for which they dig earnestly with both hands for as much as an hour (Fig. 2.3). Consequently, I had assumed that the most important dietary

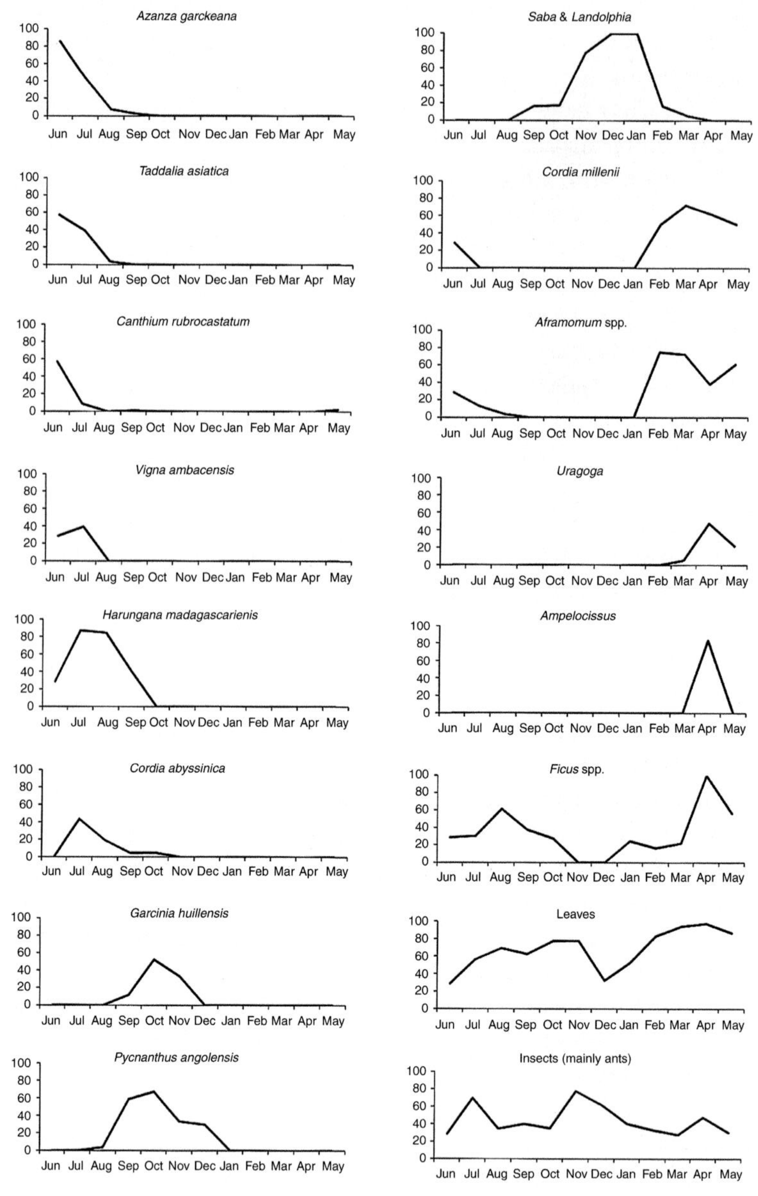

Fig. 2.2 Faecal analyses in 1975–1976 (percentage of faecal samples containing food-type by month).

difference between humans and chimpanzees was the former's use of a plant's subterranean organs (Nishida & Uehara 1983).

Recent discoveries tell us that this is not always the case. The chimpanzees of Bossou, Guinea, not only depend on cultivated fruits

Fig. 2.3 Chimpanzee digging for *Aeschynomene* root.

such as oil palm and papaya for survival, but also uproot manioc with their hands (Sugiyama & Koman 1987; Yamakoshi 1998). Furthermore, at Ugalla, the easternmost boundary of the chimpanzee distribution, where juicy fruits are hard to obtain in the harsh dry environment (Kano 1972; Itani 1979; Nishida 1989), chimpanzees use digging sticks to obtain subterranean food (Hernandez-Aguilar *et al.* 2007). Moreover, in 2001, chimpanzee fossils were discovered for the first time, in Kenya. The site was 600 km east of the current chimpanzee distribution in Uganda and also east of the Great Rift Valley (Kano 1972; Itani 1979; Nishida 1989). The teeth were 500 000 years old. Excavation suggests that no great forest has developed there since ancient times, except for riverine gallery forest. This implies that chimpanzees, at least in some parts of Africa, have a long history of foraging in savanna woodland. Taking these observations into account, it is likely that the last common ancestor (LCA) of humans and chimpanzees used subterranean resources to some extent.

The yellow baboons at Kasoje *do* eat mushrooms and dig up and munch on various kinds of subterranean resources as well. It seems that baboons have a deeper relationship with the savanna than do the chimpanzees of Mahale.

2.1.3 Differences in plant food repertoire between Gombe and Mahale

Gombe and Mahale Mountains National Parks are on the eastern shore of Lake Tanganyika, only 130 km apart. The two populations belong to the same subspecies, and the vegetation is generally similar, although Kasoje forest is more humid.

Almost 30 years ago, we compared the plant food repertoires of Gombe and Mahale (Nishida *et al.* 1983). Not surprisingly, important food items such as those listed above are eaten at both sites. Typically, we found differences only in minor foods, except for palm fruits.

Blepharis, a herb of the Acanthaceae family, grows along the lakeshore's rocky areas, and it is thistle-like, with prickly leaves. The chimpanzees of Mahale eat this leaf while grimacing and showing extreme caution (Fig. 2.4). However, the chimpanzees of Gombe do not eat it. In addition, the chimpanzees of Mahale eat the xylem of dead trees of *Ficus* (several species), *Garcinia*, *Pycnanthus*, *Pericopsis*, and *Parinari*, but at Gombe they will not give these even a second glance.

On the other hand, for the chimpanzees of Gombe, the pulp of an oil palm tree's fruit, which is abundant in fat, is the most essential part of their diet (Wrangham 1975). It is available throughout the year. They also eat the pith from the frond of the oil palm. Oil palms grow in places

Fig. 2.4 *Kasonta* eats the leaves of *Blepharis*.

where villages were present at Mahale as well, but the chimpanzees do not use a single bit of it. Perhaps the fact that humans lived in the vicinity of the oil palm groves at Mahale until recent times (the early 1980s) may explain why it is not in the apes' dietary repertoire; that is, they scarcely have had any opportunity to taste it. The villagers at Gombe were evacuated when the game reserve was designated in the 1930s, giving the apes much longer exposure to the resource.

2.1.4 Why on earth do chimpanzees congregate?

The distribution of some important fruit trees is limited to particular parts of the home range. Chimpanzees are fond of *Garcinia huillensis* fruit, a relative of mangosteen, which has a sour–sweet taste. *Garcinia* groves are found in only two sites in M-group's range, namely the southern-most hills and the north-central hills of the territory. Depending on the year, the southern-most part of the range may have an abundant harvest, while the north-central part does not, or vice versa; some years both parts have scanty harvests. When harvests are plentiful in the southern-most parts, for instance, practically every member of the group concentrates foraging in a roughly 2 km^2 region for over a month or so. At that time, if you were to search for chimpanzees in the remaining 28 km^2 area, you would be lucky to encounter a single one. Therefore, if M-group's 30 km^2 area were divided into two territories, it would be impossible for 40 members to inhabit either one over the course of a year, as in any given season, one or the other would suffer a food shortage.

On the contrary, the *Saba* fruit of the oleander family, which is a large, juicy fruit, can be found in massive amounts spread over most of the group's territory (Fig. 2.5). Since *Saba* is a woody vine, it can spread over several tall trees and forms huge food patches at Mahale. During its fruit-bearing season, a sight not to be missed is the movement of the biggest party of the year, often including every member of the group, southward or northward on the trail of these fruit (Fig. 2.6). While foraging in this mass movement, it is common for the front-line and the rear-guard of the group to be separated by over 1 km.

If large groups were to gather around a single fruit tree and start eating, the fruit would be gone in no time, because chimpanzees have such large bodies. Rather than waste a lot of energy moving from fruit tree to fruit tree, chimpanzees split into a few parties and disperse while foraging.

Fig. 2.5 Fruits of *Saba comorensis* (courtesy of Michio Nakamura).

If you track a specific individual all day, you find that chimpanzees travel for many hours in very small parties or alone (Itoh & Nishida 2007). During periods of food scarcity, individuals often travel alone, and thus it is fair to assume that solitary foraging is more efficient than group foraging. But when food supply is plentiful, why do these chimpanzees form such large parties? Congregating offers many benefits, if we put aside the lowered efficiency of fruit picking. First, congregations may be effective in preventing attacks from predators such as lion and leopard. Lions have eaten chimpanzees of M-group (see below). Most importantly, for both females and males, they can depend on one another to fend off attacks from neighbouring groups. Also, travelling alone, there is no one to groom you and to remove ectoparasites from your back. In addition to the above, hunting for colobus monkeys requires a large number of individuals.

For a male chimpanzee, congregating means closer proximity to oestrous females. A larger group also makes it easier for an individual to gain the support of allies, although at the same time it increases the chance of encountering rivals and being threatened by dominant males. Females also gain protection from harassment by violent young adult or adolescent males. Youngsters in congregations acquire a host of playmates.

Accordingly, when the foraging competition is minimal, chimpanzees who form large groups are advantaged. If a particular region has a

Fig. 2.6 From late dry season to the first half of the rainy season, M-group chimpanzees often travel long distances together.

concentration of nourishment, then there is all the more reason to congregate, since this is the most efficient harvesting method possible.

2.1.5 Travelling together

I was often surprised to see how chimpanzees travelled together. After the midday siesta, the alpha male, or any other high-ranking or older male, stands up on all fours. At this time, all of the other chimpanzees arise and pant-hoot, before starting to move rapidly *en masse*. Once, we could hardly keep up with them for 15 minutes, until they finally arrived at an ant fishing site. It appeared that, at the outset, the chimpanzees instantly understood the destination of the leading male, or perhaps they shared the common idea of visiting the same place.

One day a few adult males of M-group were sitting close by, and a young adult male, *Jilba*, touched the prime male *Kalunde* on the shoulder. *Kalunde* immediately stood up and they travelled together for five minutes or so, and then began to cooperate in hunting colobus monkeys. Such episodes occur often, and I daresay that these chimpanzees can communicate and understand quickly their companion's intentions, perhaps more quickly than do their human counterparts.

2.2 DISPERSAL OF SEEDS

2.2.1 Germination of seeds passed through the chimpanzee gut

What role does the chimpanzee play in the ecosystem? They eat leaves, flowers, seeds, buds, inner bark, and so on, and so in this respect they are destructive. However, they usually eat the parts of plants that regenerate. They do eat the seeds of the bean family, such as *Parkia*, *Jubernardia*, *Brachystegia*, *Pterocarpus*, *Baphia*, and *Vigna*, but as abundant quantities of beans grow at one time, the amount that chimpanzees can consume is limited. The apes adore the buds, blossoms, and leaves of *Pterocarpus*, and when eating them they bend and break the branches of the tree, causing the branches to die. Consequently, every *Pterocarpus* tree has a unique shape, making it easy to see the impact chimpanzees have had on them. However, I have never seen an entire tree that died from chimpanzee feeding.

Another kind of tree with branch shapes significantly altered by chimpanzees is *Pycnanthus* (a relative of the nutmeg tree). The chimpanzees of Mahale seem to consider this tree the perfect bed site. In the evening, if they enter a forest where *Pycnanthus* is abundant, I can say to myself with a sigh of relief, 'Ah good, they have found a bed for the night.' The trunk stretches straight up and then, at about 10 m above the ground, the boughs split into four or more directions. A chimpanzee sits in the centre, and by bending and drawing in the boughs, it makes a sturdy bed, making this the tree of choice for slumber.

The vital role of the chimpanzee as a 'disperser of seeds' cannot be overstressed. Since their staple food is ripe fruit, they swallow many seeds. In 1975 we began conducting faecal analysis on a regular basis. We put the faecal samples we had gathered into a metal basin, sluiced them in water, poured the liquid through a mesh screen, and, after differentiating and counting them, threw away the seeds into a large hole in the ground. A few weeks passed, then we noticed a leafy *Saba comorensis* vine and a *Cordia* sapling sprouting from the hole. Seeds that had passed through chimpanzees' guts were germinating!

Why do chimpanzees swallow the seeds? We can easily spit out the seeds of fruits cultivated by humans. However, the fruits of wild plants differ in structure from those fruits that have been domesticated through artificial selection. In their natural condition, seeds are firmly embedded in the pulp by sticky substances that make it difficult to separate pulp from seeds. For example, the ripe fruit of *Cordia millenii* is fragrant and sweet, but it is impossible to pull the pulp from the large seed, given these sticky substances. Therefore, chimpanzees swallow the whole fruit. This mechanism provides the strategy by which the plants gain reproductive benefit from seed dispersal by animals.

In 1981 and 1983 Hiroyuki Takasaki set up a makeshift farm to compare seeds he had collected from chimpanzee faeces versus seeds he had gathered from various fallen fruits on the ground. The nine kinds of fruit seeds that he planted not only germinated, but also seeds that had passed through the guts of chimpanzees, at least for *Myrianthus* and *Pycnanthus*, had significantly higher germination rates (Takasaki 1983; Takasaki & Uehara 1984). This finding suggests that these trees and chimpanzees have a close, coevolutionary relationship. This pilot experiment showed us that chimpanzees and at least some plants have evolved a partnership for seed dispersal. They exchange benefits in a trade: plants give nutrients to chimpanzees and the chimpanzees help the plants to disperse seeds (Janzen 1983).

2.2.2 Increasing plant distribution

At Mahale at least, a more suitable disperser of seeds than the chimpanzee is unthinkable. First, they eat hundreds of kinds of fruit. In addition, as they are long-distance walkers, they are bound to excrete the seeds they have swallowed in places far away from their foraging zones, even in the open land that farmers abandon after cultivation. This probably facilitates the regeneration of forest.

Furthermore, as chimpanzees visit various kinds of vegetation zones, the seeds from one vegetation zone get the chance to germinate in different vegetation types. Given the likelihood that some seeds are scattered in unsuitable places, there is a high rate of failure. However, as seeds are transported to the very limit of the chimpanzee's maximum range, plants get a chance to significantly enlarge their distribution.

Such transport of seeds to great distances is not confined to chimpanzees; birds and fruit bats also perform this task. However, chimpanzees are large-bodied, enabling them to transport smaller seeds, such as those birds and fruit bats might transport, but also larger ones. *Pycnanthus*, *Cordia millennis*, *Pseudoipondias*, *Syzigium*, *Myrianthus*, *Parinari*, *Canthium*, and so on are the chimpanzees' favourite food sources that have bulky seeds; the chimpanzees of Mahale are probably the only arboreal vertebrates there that can swallow them whole. Just by the mouth of the lakeshore in Nkala Valley, there is an isolated *Parinari* grove in the sandy soil. Usually, *Parinari* does not grow in such places. The nearest *Parinari* forest is 2 km away from this site, on the high ridge over Kasangazi Valley. I'm sure chimpanzees transported those seeds.

Chimpanzees are not the only seed transporters of the fig-like fruits with soft flesh and thousands of tiny seeds. Other animals that perform this task include the yellow-bellied bulbul, African green pigeon, Ross' touraco, and several other species of birds, as well as such arboreal mammals as the blue monkey, thick-tailed bush baby, sun squirrel, and bats; then there are those mammals that eat fruit from or near the ground, like bushbuck, blue duiker, bushpig, and so on.

I wonder why it is that all plants do not produce fruit with tiny seeds, but I guess it is because of the disadvantages faced by the smaller seeds. The nutritional reserves available for germination and development are so small that if the seed is not favoured by the ideal environment, the seedlings cannot survive. Meanwhile, the smaller number of larger seeds is compensated by a high germination rate, making them competitive. In other words, larger seeds and smaller seeds each have their own positive and negative points; advantage and disadvantage result from trade-offs.

Chimpanzees also eat hard-shelled fruit. Red-tailed monkeys occasionally eat hard-shelled fruit such as *Saba* and *Landolphia* from the Apocynaceae family, but inefficiently. They bite a hole in the fruit and then poke a finger through to the flesh to remove it, but the seed usually remains in the centre of the fruit. Even if they do eat a seed, they spit it out on the spot. Chimpanzees use their big hands to split the fruit into two halves and then consume it with their large mouths.

Here, I point out some fruits that have special coevolutionary relationships with chimpanzees. The fruit of *Voacanga lutescens* is spherical and about 8 cm in diameter (Fig. 2.7). The skin of the fruit is about 4 cm thick, while the diameter of the inner cavity containing the pulp measures only 2 cm. The rind is so thick that a five-year-old juvenile

Fig. 2.7 The fruit of *Voacanga lutescens*, with the pulp eaten by chimpanzees.

typically cannot crack it open, and even adults find it time-consuming. In spite of this, chimpanzees eagerly search for these fruits, because the pulp is remarkably sweet. For the monkeys, the shell of the fruit is too thick, but if it were thicker, even chimpanzees would find it unattractive to eat, as the pulp would be even less and eating it would be even more time-consuming. The thickness of the skin and the amount of sweet pulp seem to have reached an equilibrium that encourages chimpanzees to eat it, while discouraging monkeys from doing so due to the investment cost. Therefore, we can state a fairly robust hypothesis that the structure of the fruit is the result of coevolution with chimpanzees.

The fruit shell of *Strychnos* is extremely hard, and apart from an adult chimpanzee or baboon, no animal at Kasoje can gnaw through it. Perhaps this fruit is also the result of coevolution, but not only with baboons and chimpanzees, but also elephants and human beings, as the distribution of the tree is widespread throughout the Miombo woodland.

2.3 TASTES OF FOOD

2.3.1 Tastes of some common food items

Given that chimpanzees are our closest relatives, their taste for food should be similar to ours. I sought to test whether or not this assumption is correct. As an obvious starting point of this study, I tried some chimpanzee fruits that the local Tongwe people sometimes ate in the past. Being sweet or sweet-tart, most of them were quite acceptable. For me, the tastiest fruit chosen by the chimpanzees was *Garcinia* (the genus of the mangosteen): inside its bulbous, 3 cm diameter awaits a juicy, orange-coloured, sweet-tart flesh.

In Kasoje, over ten different species of fig occur, and the smallest of them (*Ficus urceolaris*) turns reddish-black when ripe; the sweetness and crunchy texture are exquisite, like no other fruit. Chimpanzees usually eat this species just before it is fully ripe, but in that state it is not quite as delicious. *Ficus capensis*, or the *ikubila* as it is called by the locals, is a fig that is about 4 cm in diameter. Its sweetness is inferior to the above fruits, but since it is larger-sized and grows in clusters, my colleagues in the field once made jam from it.

In the same mulberry family is *Myrianthus arboreus*, a large fruit with a refreshingly sweet tartness, which both young and old local people love. The older children climb the tree and pluck fruit while the younger children remain below to catch and gather it. One time, a group of chimpanzees showed up while the children were busy harvesting, but although there were five or six children the chimpanzees scared them and they ran away, so the chimpanzees collected an easy bounty.

Several other species of fruit – such as the mature persimmon-tasting *Uapaca kirkiana* or *Parinari curatellifolia* (with a kind of sweet flavour, reminiscent of German potatoes) and chestnut-tasting fruits from the *Canthium* genus – ripen after falling to the ground, where humans, chimpanzees, and baboons all compete for them.

Elephant grass (*Pennestum purpureum*), the tallest of the grasses, can grow as tall as 3 m, but while it is still young, chimpanzees love its soft pith. A nutritional analysis showed that it has a 20 per cent protein content, but I found its taste to be totally bland. Ken-ichi Masui said it was tasty when cooked as tempura.

The pith of arrowroot herb (*Marantochloa leucantha*) is also very bland. The stem pith of the ginger lily (*Costus afer*), with its tart flavour, and that of African ginger (*Aframomum* spp.) have a faint gingery taste to them. Chimpanzees do not eat the root stems of ginger. In Kasoje there grows the same variety of the bracken fern (*Pteridium aquilinum*) as is found almost everywhere in Japan. During the height of the rainy season, the chimpanzees eat the apex of the bracken fern but not its stem. The Japanese eat the stem after taking out the bracken fern's bitterness with ash and boiling it. The leaves of trees, from what I could tell by tasting, were for the most part edible. In particular, there were two species of *Sterculia* (family Sterculiaceae) and *Blepharis* (Acanthaceae) with prickly leaves that, when chewed up, had the smooth sliminess of okra. I hasten to add that no matter what kind of fig leaf I tried, they all smelled like spoiled milk and tasted awful.

In the dry open forest, a species of tree called *Brachystegia bussei* (a type of miombo, a member of the family Caesalpiniaceae) grows. K-group enjoyed gnawing on the inner layer of its bark during the middle of the rainy season. At this time, under the *Brachystegia* trees, one sees bark pieces with many chimpanzee teeth marks that cover the inner surface of the bark. If you lick it ever so gently, it tastes sweet. In North America there is a tree called the sugar maple, which is grown for the sole purpose of harvesting sugar. During the rainy season, the cambium layer of the *Brachystegia* may also store an abundance of sugar (Nishida 1976).

Chimpanzees eat about ten kinds of tree gum, such as that from *Terminalia* and *Acacia*. One day, as it was approaching evening, the group began to pick up speed. Just when I thought I had lost sight of them, from the front of the line a pant-hoot vocalisation was heard: was there meat to eat? Panting and out of breath, I ran up to check on the commotion. There was a large crowd of apes sitting in a silk tree (*Albizzia glaberrima*), about 3–5 m up the trunk and in a very excited state; they were eating something, but it was not meat! A stream of gum was oozing from the trunk, and they were scooping it up and sucking it down (Fig. 2.8). Much to my dismay, I never got a chance to taste this delicacy that brought the chimpanzees to such great excitement.

2.3.2 Tastes chimpanzees share with us

The diet of chimpanzees can also include human food. In many parts of Africa, wild chimpanzees cause damage to such crops as papaya, grapefruit, orange, lemon, mango, guava, oil palm, the stalks of banana, maize and sugarcane, manioc, and sweet potato (Nissen 1931; Kortlandt 1962, 1967; Goodall 1968; Nishida 1968; Albrecht & Dunnett 1971; Kano 1972; Sabater Pi 1979; Nishida & Uehara 1983; Takahata *et al.* 1984; Takasaki 1983; Sugiyama & Koman 1987; McGrew 1992; Yamakoshi 1998). Additional evidence for shared taste comes from neuro-physiological studies of the chimpanzee's sense of taste. For example, taste nerve fibres in the chimpanzee fall distinctly into categories congruent with human taste qualities. The chimpanzee's tongue may have areas with partial taste specialisation that corresponds to the taste nerve fibres that respond mainly to bitter, sweet, or salty substances (Hellekant &

Fig. 2.8 *Ntologi* eats the resin of *Albizia glaberrima*.

Ninomiya 1991). The PTC (phenylthiocarbamide) taste polymorphism exists in chimpanzees, and the frequency of 'tasters' is as high as that of humans (Kalmus 1970).

2.3.3 Trying to eat everything chimpanzees eat

As mentioned above, I sampled only food items that appeared to be edible, besides those that my Tongwe assistants taught me were edible. This is because I once had a terrible experience when trying a new food. It was the fruit of *Pycnanthus angolensis*, a relative of nutmeg used as a fragrance in Southeast Asia. Inside the hard fruit shell there is a large seed, which is the perfect equivalent of nutmeg. Then there is a flimsy red aril covering the seed. This corresponds to the part of nutmeg that is called 'mace', and chimpanzees only eat this part (Fig. 2.9). When I put the aril into my mouth, the bitter and unpleasant taste so

Fig. 2.9 The bitter aril of *Pycnanthus angolensis*.

overwhelmed me that I immediately spat it out. Watching this, my Tongwe assistants burst into laugher. I wonder if, because of this bitterness, chimpanzees swallow it, seed and all? Anyway, this aril of *Pycnanthus* had a taste far too awful for words, and so for a long time I gave up trying to taste chimpanzee foods, aside from those that the Tongwe people told me were alright to eat.

One problem with these observations, however, is the lack of systematic sampling and the possibility of unconscious selection of apparently edible food items. Another difficulty is that researchers cannot always identify foods that grow high in the canopy, or those that are eaten only rarely. If the subjects of a study are unhabituated, then it is almost impossible to sample all of the food items being consumed. A better approach is based on a less-biased sampling of food items eaten by habituated individuals.

In 1991 I began to investigate the chimpanzee's sense of taste by directly sampling chimpanzee foods. I carried out five short field studies spanning a total of 13 months, consisting of 2–3 months each time in 1991, 1992, 1994, 1995, and 1997. I followed 6–9 adult male targets, recorded their eating bouts, and tasted the same food items eaten by the apes. In each study period, I recorded all of their eating, and made every effort to taste the same food items eaten by the target. However, this was not always possible. When the target was eating food from a tiny patch – the young leaves of a small shrub, for example – the ape often consumed all of the edible items, unless I snatched the last piece of food. This inevitably resulted in aggression towards me. When the chimpanzee was eating high above the ground, I could only pick up and taste fallen items or abandon any effort to taste that food. However, these cases were exceptional; most of the time I had ample opportunities to obtain and taste the same food eaten by the target male.

Taste was divided into 14 categories on the basis of five tastes (sweet, salty, sour, bitter, astringent) and their combinations. 'Sweet, salty, sour, and bitter' are based on four basic receptors (Sutherland 1987). 'Astringent' is the taste sensed when tannin-rich food such as tea or persimmon is ingested.

I tasted each food item at least twice over several field seasons, but my judgements were consistent most of the time. One more category was 'insipid' (that is, having none of the above tastes).

I tasted 95 species (or 114 food items) during the study period. These comprised 48 per cent of the 200 species or 34 per cent of the 336 plant food items recorded for the chimpanzees of Mahale (Nishida & Uehara 1983; and my unpublished data). Nine kinds of fruit in both ripe and unripe states are eaten by chimpanzees. Since ripe fruits differ in taste from unripe ones, these were recorded as different items. Thus, I tabulated the tastes of 123 food items (Nishida *et al.* 1999).

The tastes of the parts eaten are summarised in Table 2.2, in which tastes of the same kind that differ only in degree are arranged close to each other. For example, LS (slightly sweet), SS (moderately sweet), and VS (very sweet) are lumped as the same taste, 'sweet'. The 'salty' taste was completely lacking in the chimpanzee foods sampled.

Surprisingly, foods were most often characterised as 'insipid', followed by 'sweet', and then 'bitter'. Ripe fruits, the single most important food item for chimpanzees, were either sweet or sweet and sour, while unripe fruits were often astringent or bitter. Pleasant and palatable foods, that is, mostly insipid, lightly bitter, lightly sour, sour and sweet, lightly sweet, moderately sweet, and very sweet comprise 69 per cent of food items.

To infer the gustatory world of wild chimpanzees from the sense of taste of a human observer, we must consider the duration of time spent eating each food item. Times spent eating in 1994 and 1995 were analysed, and each food item was divided into nine major taste categories. Fig. 2.10 shows time spent consuming food items from each taste category. The two study periods differ by season (June–August in 1994, September–November in 1995). In 1994, when the sweet fruits of *Afrosersalisia cerasifera* (Sapotaceae) predominated, 'sweet' accounted for 54 per cent of the time; in 1995, when sour-sweet fruits of

Table 2.2 *Summary of the tastes of chimpanzee plant food at Mahale (Nishida* et al. *2000)*

	Taste category*														
Part eaten	IN	LB	BB	VB	SB	OB	LO	OO	SO	LS	SS	VS	SA	LA	AA	Total
Pith	7	3		1			3	1								15
Leaf	15	11	2		1		1	2						1	1	34
Fruit	4		1		4	1		2	8	13	11	5	5	2	2	58
Resin	1															1
Seed	2	2		1												5
Gall	2															2
Blossom	2				1			1								4
Bark										1						1
Petiole	1															1
Endocarp	1															1
Wood							1									1
Total	35	16	3	2	6	1	5	6	8	14	11	5	5	3	3	123

* IN: insipid, tasteless; LB: slightly bitter; BB: moderately bitter; VB: very bitter; SB: sweet and bitter; OB: sour and bitter; LO: slightly sour; OO: moderately sour; SO: sour and sweet; LS: slightly sweet; SS: sweet; VS: very sweet; SA: sweet and astringent; LA: slightly astringent; AA: moderately astringent

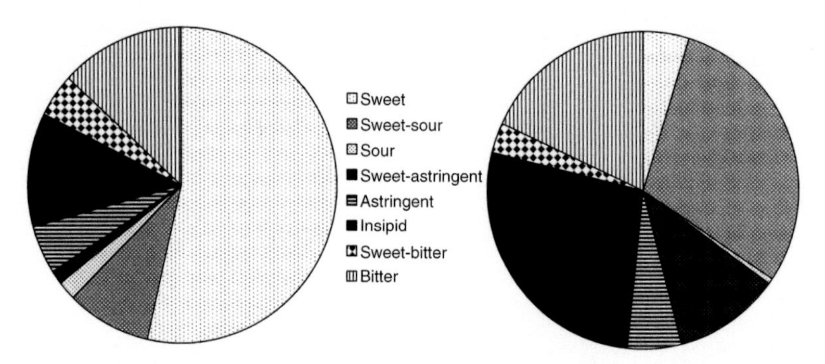

Fig. 2.10 The 'taste world' of wild chimpanzees. Illustrated by time spent feeding on food items having one of eight kinds of tastes: left, 1994; right, 1995 (from Nishida *et al.* 1999).

Saba florida (Apocynaceae), bitter arils of *Pycnanthus angolensis* (Myristicaceae), and sweet-astringent fruits of *Pseudospondias microcarpa* were available, the 'sweet' category represented only 4 per cent.

Thus, the *umwelt* of the chimpanzee from the perspective of sense of taste changes remarkably over the year. However, chimpanzee food was deemed acceptable to a human observer most of the time in both periods. Foods lacking the tastes of astringent or bitter (i.e. sweet, sour-sweet, sour, insipid) represented 76 per cent of foods sampled in 1994 and 57 per cent in 1995. This rate excludes slightly bitter foods that tasted pleasant to me, as well as sweet-astringent fruits that local people were happy to eat.

2.3.4 Food selection strategies of the African ape clade

Many piths and leaves have no taste or are only slightly bitter, and most chimpanzee food is edible. I have also sampled various foods of Japanese macaques in Chiba Prefecture, and some of them (e.g. bean of *Caesalpinia sepiaria*, Leguminosae, and fruit of *Cinnamomum camphora*, Lauraceae) were so bitter that I will never forget the experience. Hayashi mentioned a very toxic plant (*Coriaria japonica*, Coriaceae) as one of the fruits eaten by Japanese macaques at Mt. Hakusan. Murata and Hazama mentioned a poisonous, perennial herb (*Macleya cordata*, Papaveraceae) as one of the favourite food plants of the Japanese macaques of Arashiyama, Kyoto (Murata & Hazama 1968, p. 12). Kummer (1995) tasted the soft bases of the leaves of *Sansevieria*, eaten by the hamadryas baboon, and these tasted 'repulsively bitter' to him. Although a systematic comparison of the diets of chimpanzees and Old World monkeys is lacking, one can hypothesise that these two lines of primates have evolved remarkably different strategies of food selection. This agrees with Napier's suggestion that ancestral catarrhine primates originally diverged into frugivorous hominoids and folivorous cercopithecoids (Napier & Napier 1967, p. 24). This was later elaborated by Andrews and Aiello, who concluded that the ancestral catarrhine dietary condition was frugivory – and retained by the hominoids – while the ancestral condition of the cercopithecoids was folivory (Andrews & Aiello 1984).

The ancestors of Old World monkeys, as folivorous animals, have evolved elaborate digestive mechanisms for the detoxification of secondary compounds such as alkaloids.[1] In contrast, hominoids have not evolved refined detoxification systems, but instead have developed the habit of more strictly discriminating the potential for toxicity among plant foods. By comparing the plant diets of chimpanzees and sympatric cercopithecine monkeys, Wrangham and colleagues (1998) showed that monkeys had absolutely higher intakes of anti-feedants than chimpanzees. Gorillas also seem to avoid bitter food. About 16 per cent of the foods of the gorillas of Rio Muni were bitter, but most of them were only slightly bitter, while 19 per cent of their foods were insipid (Sabater Pi 1977). Although gorillas eat much vegetation that is high in tannins, very few foods seem to contain alkaloids (Calvert 1985; Rogers *et al.* 1990). This shows that at least one species of the African great apes refuses to ingest poisonous materials. The complicated feeding techniques for eating vegetation, such as leaves and piths of herbs, by gorillas (Byrne & Byrne 1993) and chimpanzees (my observations) reflect the selection of non-poisonous materials that are often protected by thorns. This tendency to select non-bitter food may be the biological basis for the hominid invention of food-bleaching and detoxification,[2] and finally the use of fire and other cooking methods (Hladik & Simmen 1996). Descendants of colobines would never entertain the notion of cooking!

Although most of the chimpanzee foods had pleasant tastes and were thus palatable to a human observer, some were very bitter,

[1] See, e.g. Waterman (1984) for a review of the relationship between food processing and plant chemistry in non-human primates.

[2] See Leopold & Ardrey (1972); Stahl (1984); and Johns (1989) for human food selection.

suggesting higher tolerance to bitter substances in apes than in humans. The chimpanzee has a threshold of bitter taste (in terms of quinine hydrochloride) that is four times higher than that of humans (Hladik 1981). Moreover, in the rain forest of Gabon, where 14 per cent of the 382 plant species reacted positively to the alkaloid test, the chimpanzee includes in its diet a similar proportion (15 per cent) of plants likely to have a high alkaloid content (Calvert 1985; Rogers *et al.* 1990). These results are congruent with my observation because the frequency (actually duration) of eating bitter items is very low.

My sample included ten fruits, seeds, or blossoms that tasted bitter. Chimpanzees seemed to eat these bitter foods because of their sugar, lipid, or protein content, and it is likely that they accepted a trade-off between nutrition and toxicity. The most unpalatable item tested, the aril of *Pycnanthus angolensis*, has an extraordinarily high lipid content (74.5 per cent of dry weight). Thus, it is likely that chimpanzees made a trade-off between high caloric value versus toxic materials (Table 2.3).

Another source of bitter food tolerance may result from their biological activity. A physiological function such as purging intestinal parasites may be sufficiently important to overcome repulsive tastes (see Section 2.4).

The chimpanzee's food selection strategy may be to: (1) prefer non-poisonous alkaloid-free foods; (2) ingest astringent (i.e. tannin-rich) foods when they are also sweet; and (3) consume some foods containing alkaloids when they have some nutritional or possibly medicinal value.

After human ancestors diverged from the LCA with *Pan*, they subsisted in more open environments such as savanna woodland. Since present savanna-living human populations show higher taste sensitivity than their forest-living counterparts (Hladik 1981), perhaps humans have evolved greater taste sensitivity to, in particular, bitterness than have chimpanzees. However, humans retain the habit of ingesting bitter foods such as caffeine and cassava (Dufour 1993). Perhaps it is a common feature for both man and chimpanzee to eat some bitter foods, although they tend to select non-bitter food if it is available. Human taste sensitivity is influenced by cultural transmission (Hladik 1993; Hladik & Simmen 1996), as is also likely for apes, but modern humans may retain some of the ancient sense of taste inherited from their hominoid ancestors.

Table 2.3 *Nutritional analyses of some important plant foods of chimpanzees at Mahale (Nishida* et al. *2000)*

Plant name	Part eaten	Water	Crude protein	Crude lipid	Crude ash	Carbohydrate	Energy (kcal per 100 g)
Aframomum mala	Pith	10.6	14.5	2.2	19.2	53.5	292
Afrosersalisia cerasifera	Fruit	12.3	4.8	2.8	3.1	77.0	352
*Baphia capparidifolia**	Leaf	No data	41.3	No data	5.8	10.5	No data
*Baphia capparidifolia**	Flower	No data	33.0	1.6	6.0	14.6	No data
Canthium rubrocostatum	Fruit	9.2	9.5	26.5	1.4	53.4	479
Cordia millenii	Flower	11.0	23.1	2.5	9.4	54.0	295
Diplorhynchus condylocarpon	Seed	8.9	15.0	21.6	3.5	51.0	458
Ficus exasperata	Fruit	11.0	4.2	4.4	2.9	77.5	366
Ficus exasperata urabe	Leaf	14.8	15.3	2.3	11.2	56.4	234
Ficus sp. (mjimo)	Leaf	9.9	26.3	4.4	10.6	48.8	340
Ficus sp. (mtoligana)	Fruit	12.8	5.9	4.6	3.9	72.8	356
*Ficus urceolaris**	Fruit	13.7	5.3	6.3	8.0	66.7	345
*Ficus urceolaris**	Fruit	18.7	9.8	7.4	7.0	57.1	334
*Ficus vallis-choudae**	Fruit	14.3	4.1	5.4	6.6	69.6	343
*Ficus vallis-choudae**	Fruit	15.4	7.5	3.8	7.3	66.0	281
Landolphia owariensis	Pith	11.3	9.9	4.8	6.3	67.7	316
Milicia excelsa	Flower	12.3	16.3	2.2	9.5	59.7	283
Mlangi	Flower	14.4	19.6	1.8	7.3	56.9	294
Parinari curatellifolia	Fruit	17.1	2.6	0.6	3.2	76.5	322

Table 2.3 (*cont.*)

Plant name	Part eaten	Water	Crude protein	Crude lipid	Crude ash	Carbohydrate	Energy (kcal per 100 g)
Pennisetum purpureum	Pith	13.1	19.1	1.5	16.3	50.0	290
Pennisetum purpureum urabe	Pith	12.1	18.2	1.4	14.1	54.2	219
*Pseudospondias microcarpa**	Fruit	18.0	6.8	9.2	5.5	60.5	330
*Pseudospondias microcarpa**	Fruit	20.3	7.1	0.6	5.4	66.6	300
Psychotria peduncularis	Fruit	18.9	5.4	3.3	6.1	66.3	290
Pterocarpus tinctorius	Leaf	11.6	26.1	2.3	5.4	54.6	344
Pycnanthus angolensis	Fruit	3.5	4.4	74.5	1.3	16.3	753
Saba comorensis	Fruit	23.0	4.7	5.3	3.8	63.2	319
Syzygium guineense	Fruit	21.3	2.9	2.1	2.5	71.2	315
Uvaria angolensis	Fruit	15.8	5.1	0.9	3.5	74.7	327
Vernonia amygdalina	Pith	13.9	19.1	1.3	14.0	51.7	247

* Two samples were analysed.

2.4 LEAF SWALLOWING AND BITTER PITH INGESTION

2.4.1 The chimpanzee's herbal medicine? The riddle of the wild aster leaf

When a wild chimpanzee eats tree or vine leaves, they either strip the stem by holding it in one hand and then put into their mouth the leaves that they have stripped and eat them, or they strip the stem directly with their lips and then eat the leaves. Whichever way they choose, they can eat 20–30 leaves per minute, thoroughly chewing and crushing the leaves before swallowing them.

A few kinds of greens, especially from the legume family, supply the chimpanzee's most vital sources of protein: *Pterocarpus tinctorius*, *Baphia capparidifolia*, and, as mentioned above, the piths of elephant grasses.

However, the *Aspilia* leaf (a relative of the aster) and at least six other kinds of leaves, are swallowed slowly by the chimpanzee, one-by-one as if taking medicine (Fig. 2.11). This particular manner of ingestion often takes place in the early morning, between 6:30 a.m. and 9:00 a.m., although it may occur at any time, at least at Mahale (Wrangham & Nishida 1983; Takasaki & Hunt 1987).

Fig. 2.11 A male chimpanzee swallows the leaf of *Commelina*.

In addition to the chimpanzee not chewing the *Aspilia* leaf and so not releasing nutrients, only about five or six leaves are swallowed per minute; it makes no sense to say that the ingestion yields nutritional value. So why on earth does the ape eat them in this manner?

During his visit to Mahale in 1971, Richard Wrangham, who first noticed this strange behaviour at Gombe, told me that chimpanzees swallowing *Aspilia* might be similar to humans drinking coffee. When I started faecal analyses of chimpanzees as a long-term project in 1975, I found some leaves that had not been digested, still perfectly intact. These leaves proved to be from *Aspilia*.

Following Richard's first suggestion that it might be a kind of stimulant, we thought it might be medicinal, and so I looked up *Aspilia* in a book of ethnomedicine, or traditional pharmacology. I found that African people use plants from the genus *Aspilia* to treat cystitis, gonorrhoea, abdominal pain and other ailments, and for ridding themselves of intestinal parasites (Kokwaro 1976). When I reported this to Richard, he asked us to collect as many *Aspilia* leaves as possible, and Shigeo Uehara and I complied with his request. He forwarded the samples to a specialist at the University of California for biochemical analysis. The notion that chimpanzees likely were using the *Aspilia* leaf as medicine became more plausible when the California professor conducting this analysis announced that he had extracted the antibiotic, thiarubrine A (Rodriguez *et al.* 1985). In small doses, this compound is effective for killing bacteria, fungi, nematodes, and so on. This caused an uproar in the international academic world, as well as in the mass media, such as *Time* magazine.

Alas, the experimental result proved to be based on a contamination of leaves and roots! A researcher at a Canadian university, in a later study of the leaf, found that there were only trace amounts of thiarubrine A present. In the younger leaves, there is a variety of diterpenes (the bitterness found in acidic fruits) called grandiflorenic acid, which has the unique ability of inducing uterine contractions. It is also hypothesised that females use it as a kind of birth control drug (Page *et al.* 1992), although this does not explain why males eat the leaves. Whatever the medicinal components may be, apparently healthy chimpanzees ingest the *Aspilia* leaf, so there is a high probability that chimpanzees also use it for disinfection of intestinal parasites. According to faecal analysis, *Aspilia* most often is swallowed during mid-rainy season, around January or February. Also, stool tests run by Masato Kawabata (now Professor at Kobe University) many years ago indicated that chimpanzees' parasite infection rate was highest during this period (Kawabata & Nishida 1991).

In the same manner as with *Aspilia*, *Trema* (from the elm family), *Lippia* (a variety of fig), and the dayflower (*Commelina*) are swallowed slowly, in the early morning, suggesting that they too contain medicinal components (Newton & Nishida 1991). All of these leaves swallowed whole by the chimpanzees have hard hairs sprouting from them, giving a coarse surface. This compels us to consider the possibility that, in one way or another, the leaves' physical properties are used. Michael Huffman of Kyoto University proposed the hypothesis that hairy leaves catch and trap intestinal parasites and thus help to purge them out of the intestines (Huffman 1997).

Moreover, whenever chimpanzees eat *Aspilia* leaves in the afternoon, not only do they chew them, but also they eat them quickly and in large amounts. In short, it appears that in the afternoon the leaves are eaten as food. Currently, I think that chimpanzees usually eat *Aspilia* leaves simply as food, but sometimes they use them for medicinal purposes. No satisfying answer to this riddle has yet been found.

2.4.2 Bitter *Vernonia* leaves

In 1987 Michael Huffman made an intriguing sighting (Huffman & Seifu 1989). A sick female chimpanzee, *Chausiku*, was eating the terribly bitter ironweed plant of the Asteraceae family. Neither *Wantendele*, the female who was travelling together with *Chausiku*, nor her ten-year-old son, *Masudi*, ate the *Vernonia amygdalina* pith. Michael thought that as the sick chimpanzee ate it, but the healthy ones did not, the pith might contain a certain biologically active agent working as medicine. He confirmed that *Chausiku* recovered her health within 24 hours of eating the *Vernonia* piths.

From the bitter pith of *Vernonia amygdalina*, Koichi Koshimizu, Hajime Ohigashi, and their colleagues in Kyoto University's Agricultural Department extracted bitter-tasting compounds of biological or physiological significance, e.g. sesquiterpene lactone (vernodalin, vernolide, etc.) and steroid glucosides (vernonioside A1–A4, B1–B2) (Jisaka *et al.* 1993; Koshimizu *et al.* 1993; Ohigashi *et al.* 1994; Huffman 1997).

The chimpanzees may use the piths of *Vernonia* to control illness, but some questions remain regarding this proposal. First of all, I have seen many apparently healthy chimpanzees, including *Chausiku* in

1975, often eat the piths of *Vernonia*. In my observations, *all* chimpanzees that I saw eating *Vernonia* leaves were apparently in good health. Second, it has recently been found that the young piths of *Vernonia* contain much protein, at about 20 per cent of dry weight (Table 2.3). Therefore, it is likely that chimpanzees eat these piths for nutritional purposes. If the energy needed to detoxify the bitter factors is less than the nutritional value that the pith provides, it can reasonably serve as a food even if it tastes bitter. Nevertheless, the piths could be consumed as both food and medicine, thus playing a role such as the control of intestinal parasites. If this is the case, chimpanzees in good health might well eat these piths in great amounts.

Anecdotal reports continue to suggest that baboons, muriquis, howler monkeys, and brown bears, among others, use plants for medicinal purposes (Strier 1998). If chimpanzees and other animals really are practising pharmacy and using natural medicine, this potentially could have a great influence on nature conservation. This might mean that humans could use such observations for screening biologically active agents and so enable us to find medicine for humans. If this actually happened, those who have no understanding of the importance of wildlife conservation might snap out of their complacency and come to appreciate the great importance of promoting it. In the natural world, the secrets yet undiscovered by humanity are numerous. However, definitive answers can come only from the long-term health-monitoring of chimpanzees and chemical analyses of the substances ingested by sick chimpanzees. Leaves of *Aspilia* and the like and the piths of *Vernonia* were eaten in large amounts by apparently healthy individuals in all of my observations. Before advocating 'chimpanzee pharmacology' or 'self-medication', scientists must carry out more extensive investigations.

2.5 HUNTING

2.5.1 Going on a colobus hunt

Looking at wild chimpanzees, I find the most fascinating thing about them to be their hunting of colobus monkeys. During my first 15 years in the field, I mainly watched K-group, which was then composed of only 20 or so members, and I hardly ever saw them hunt colobus. Colobus hunting requires a large party, which K-group could not muster. I observed K-group chimpanzees grab fawns of blue duiker and bushbuck, bushpiglets, and fledglings of francolin, guinea fowl, and other small birds. This was more like 'gathering' than hunting of animals, although the apes occasionally caught juvenile red-tailed monkeys (Nishida *et al.* 1979).

In 1981 I shifted my major observations to M-group. At the time, my research centred on social relationships among newly immigrated and resident females, something I then knew next to nothing about. Day in and day out, I targeted one female and her dependent offspring and followed them everywhere. Females mostly pass the time in small groups, foraging or nursing, and aside from the occasional grooming party, rarely pant-hoot. During one of these predictable observations, I suddenly heard noises of uproar and turmoil, pant-hoots and screaming in the distance. Pant-hoots are given in the heat of the moment, signalling that something has been caught to feast upon.

The chimpanzees started getting excited, madly chasing one another, appearing frenzied. 'Hunting!' had obviously happened, but as long as the day's targeted female did not go to the place of the kill, I had no idea of who caught the prey or even what had been caught. Unless I stopped accumulating data for the primary purpose of my research, which was social relationships among females, I would have to give up the idea of observing the hunt. The exasperation I felt then was nothing to be ashamed of!

The red colobus mainly feeds on young leaves and is an arboreal monkey that travels in troops of 20–40 (Nishida 1972b; Ihobe 2002; Struhsaker 1975; Stanford 1998). They are often hunted by the apes when they are clumped together in the tops of giant silk trees (*Albizia glaberrima*). Ten chimpanzees may climb the giant tree, and then they will apparently chase down the colobus in free pursuit, but it is clear that the chimpanzees anticipate the directions their hunting partners take and use one another's positioning in the hunt. The hunters pursue, kick and rattle branches, and try to corner the colobus. When hunted down, the mother colobus becomes frightened, and the chimpanzee reaches and snatches away their young, without the mother raising even a finger of resistance.

The captured prey are generally infant, juvenile, adolescent, and adult female colobus, with the largest males of the troop never being captured. It is not uncommon for a chimpanzee to let out a ghastly scream and hit the ground, fleeing after facing a fierce counterattack from an adult male colobus. Recently, adult male colobus have been seen to descend to the ground to chase away chimpanzees, and occasionally even human researchers. Infant, juvenile, and younger adolescent colobus are caught mostly in the treetops, but adult and large

Fig. 2.12 A red colobus leaping.

adolescent monkeys are caught more often on the ground, after being pursued, jumping, falling, and hitting the ground (Fig. 2.12).

Two types of colobus hunting have been recorded: when the kill is eaten in private, and when a few members gather for meat-feeding in the treetops or on the ground. Since the colobus has razor-sharp canine teeth, the hunter could be severely injured if he does not instantly kill a captured monkey. There are various ways for a chimpanzee to slay a colobus: usually, the hunter grabs it by the tail or leg and violently beats it against the ground, dragging it by the tail and beating its head against a buttress root, vine, or rock until it is nearly dead, or he pins its body to the ground and chews open its entrails, yanking them from the abdomen. When M-group's alpha male, *Ntologi*, had a pinned-down bushpig that must have weighed 15 kg, he delivered rapid elbow blows to its face to kill it. When the prey is an infant or juvenile, the hunter kills it by biting its face or head.

When a hunted-down adult colobus falls to the ground, it suffers from concussion and sits on the ground or seeks refuge in gaps between rocks or thickly entangled woody vines (Fig. 2.13). But if the monkey is still active, it is often the chimpanzee who suffers in trying to kill it. The colobus tries to bite and nip in self-defence, so the chimpanzee has to be careful not to lose a finger! Even in large numbers, it sometimes takes chimpanzee hunters over an hour to slay a colobus. Once, an adult colobus with an injured leg took refuge in a

Fig. 2.13 A red colobus sitting on the ground.

hole in the ground. It looked like easy prey, but after an hour the chimpanzees gave up on the kill and left the scene.

Adolescent male and young adult male chimpanzees often carry on the chase in the treetops, while middle-aged and older individuals stay at the foot of the tree from the start, tending to assume the 'good luck charm' position. This distribution of hunters looks like a division of labour between the 'beater' and the 'captor', but it does not look like one hunter deliberately chases a colobus in the direction of another hunter. Instead, it appears that every chimpanzee wants to catch the monkey *himself*. Moreover, the individual who makes the catch does not divvy up the meat with the 'beater'. Therefore, we cannot call this a true division of labour or cooperative hunting, although it is a form of group hunting. Therefore, I call this hunting style 'simultaneous individual hunting' (Nishida 1981; Takahata *et al.* 1984). Hunting at Gombe resembles that at Mahale (Busse 1978). Christophe Boesch, the Swiss primatologist studying the chimpanzees of the Taï Forest, Ivory Coast, stated that he could discriminate among hunter, beater, and onlooker and that the hunters and beaters collaborated (Boesch 1994). However, he did not provide the crucial data on whether or not hunters shared meat only with beaters but not with onlookers. We have found it impossible to distinguish hunters definitively from beaters or onlookers, as onlookers switch to being hunters or beaters, while hunters become onlookers when they get tired.

2.5.2 Meat-eating traditions

In some cases, 'catching' or 'grabbing' would be a more accurate description than 'hunting', as I mentioned above, regarding K-group's hunting style. When a blue duiker's young is taken by surprise, for example, it freezes in place and thus becomes easy prey. Bushpigs usually build nests in hay, ginger, or the like and stash their piglets there in October, at the beginning of the rainy season. All a chimpanzee has to do is reach in and snatch the piglet. However, if the parent bushpig is nearby, it can be an extremely dangerous task. In the early 1990s, an adult male chimpanzee, *Jilba*, had his left arm muscle slashed to the bone; our Tongwe assistants told us that it was unmistakably the work of a bushpig.

One time, a juvenile chimpanzee came across a sick red-tailed monkey. An adult male chimpanzee, *Bakali*, seeing the juvenile approaching the sluggish red-tailed monkey, seized the monkey in the blink of an eye.

When feasting on meat, chimpanzees always eat live or dead leaves of trees or vines at the same time. This is called 'wadging'. It does not serve the same purpose as a salad! They use leaves, such as the green leaves of *Saba comorensis*, which they normally do not eat, and often fill their mouths with dried leaves. Therefore, like Asian konjak root, or devil's tongue, this wadging may augment the amount of meat, prolonging the pleasure of a feast (Teleki 1973), or accent the cuisine. Chimpanzees eat muscle, entrails, bones, and skin.

2.5.3 Individual differences

Males do more hunting than females (McGrew 1979; Uehara 1986), but even among male chimpanzees, there are virtuoso hunters and those who are not so good at hunting. The hunter extraordinaire in the 1990s was a 16-year-old male named *Toshibo*. He was then low in social rank, and stronger males robbed him of his kills, so when he did make a kill, he ran with it and dropped out of sight from his associates.

According to our combined data (Hosaka *et al.* 2001) from 1991 to 1995, nine of the eleven top hunters were adult or adolescent males and two were adult or adolescent females. *Kalunde* was the top hunter (15 captures) followed by *Toshibo* (13), *Ntologi* (9), *Alofu* (8), and *Dogura* (7). All of them occupied the alpha or second-ranking status during or after this period.

How efficient is hunting by chimpanzees? The record number of kills in one hunt stands at 11. During 1991–1995, the average number of kills was 1.4 in a single hunt, when a different hunting event occurred after at least a one-hour interval (Hosaka *et al.* 2001).

Kazuhiko Hosaka and colleagues (2001) estimated that the yearly predation frequency on red colobus monkeys, estimated as 12 × 365 × the number of kills in observed hunts/total observation time, was 153 (1991 and 1993 data). As these 153 prey include babies, and if we estimated 3 kg per kill as an average prey weight, the chimpanzees obtained 459 kg of meat per year. If this amount were divided equally among 50 chimpanzees (which is the mean number of adult and adolescent members of M-group (Nishida *et al.* 2003) from 1991 to 1995), then each would receive 9.2 kg of hunted meat. Let us compare this figure with the performance of modern human hunter-gatherers. For example, if both men and women of the Bushmen of Southern Africa ate equal amounts of meat, each individual would obtain 82, 93, or 108 kg (Tanaka 1976; Lee 1979; Silberbauer 1981). Accordingly, the meat consumption of chimpanzees is only 10 per cent that of these particular human hunter-gatherers. If we consider that the Bushmen are a hunter-gather group that are among those depending least on animal protein (imagine the Inuits, who depend on seal meat for 90 per cent of their subsistence!), the human meat-eating tendency is much greater than that of chimpanzees.

But the meat is not shared evenly, and some individuals receive a king's share. If adult males eat 88 per cent versus adult females' 12 per cent of the meat, as suggested by Boesch and Boesch-Acherman (2000), adult males of M-group eat about 50 kg of meat per year (as the average number of adult males in 1991–1995 was eight; Nishida *et al.* 2003). This annual figure compares to 27 kg for Gombe (Wrangham & Riss 1990) and 68 kg for Taï (Boesch & Boesch-Achermann 2000). If this estimate is accurate, then little difference exists in meat consumption between some Southern African hunter-gatherers and adult male chimpanzees.

The Mahale chimpanzees' prey repertoire, aside from colobus, consists of other non-human primates such as red-tailed monkeys, blue monkeys, vervet monkeys, thick-tailed bushbabies and baby chimpanzees (i.e. cannibalism), as well as ungulates such as bushpigs, warthogs, bushbucks and blue duikers; rodents such as red-bellied sun-squirrels, African giant rats, yellow-spotted dassies; and other mammals such as civets. However, there is no evidence that they eat or scavenge on leopard, aardvark, or lion. As for birds, the fledglings and eggs of francolin and guinea fowl are much sought after in the dry season.

There has been one observation each of chimpanzees grabbing and eating fledglings of weaverbird and paradise flycatcher. In areas where villagers live, chimpanzees chase and kill chickens and occasionally steal their eggs.

In the mid-dry season, a kind of migratory bird of prey, the black kite (*Milvus migrans*), is often found resting from fatigue on the rocks of larger rivers, such as the Kasiha. It is easy for a chimpanzee to capture them. Chimpanzees often grab them, but they seem not to like to eat them. Sometimes they nibble a bit on the body, or some males, such as *Pim*, flail the carcass against rocks and repeatedly toss it about. Juveniles prefer to hold and carry the body rather than eat it. M-group chimpanzees do not accept on their menu those animal species that they rarely encounter.

2.5.4 Tool use for prodding small mammals

Chimpanzees use probes to inspect things inside cavities, where food may be hidden or danger may lurk. Another tool used, of a different type, is a stout and thick branch. I once saw an adult female of K-group stir with a probe in the hollow of a tree, before using a sturdier branch, but nothing seemed to be produced by this effort. At the time, I thought she wanted to rouse small animals such as insects or reptiles from the hole and then grab them. I also once watched *Chausiku* insert a long stout stick into a bees' hive, apparently in an attempt to get something, but many bees flew out of it, and she ran away without gaining anything.

However, in 1991, Mike Huffman saw a 12-year-old female, *Tula*, stirring around in the hollow of a tree using the above technique, and a squirrel ran out; *Tula* grabbed it and ate it immediately (Huffman & Kalunde 1993). In 1995 Noriko Itoh witnessed an eight-year-old female, *Maggie*, catch a squirrel with a stout branch (Nakamura & Itoh 2008). *Tula* was *Maggie*'s foster mother (see Chapter 3), so *Maggie* may have learned this tool-use technique from *Tula*. In 2004 Michio Nakamura observed another animal-expelling technique. Adult males forcefully pushed stout branches into rock crevices. Although they did not succeed in obtaining prey, it was the place in which hyraxes had been residing for many years (Nakamura & Itoh 2008). We occasionally observed M-group chimpanzees carrying a small hole-dwelling mammal, such as a thick-tailed bushbaby (*Otolemur crassicaudatus*) or a lesser bushbaby (*Galago senegalensis*). Although we have not seen hunting behaviour, it is likely that the chimpanzees expelled the mammal with a stick and then grabbed it by hand. Thus, the frequency of expelling and grabbing small mammals from tree holes or rock crevices is probably more common at Mahale than was imagined (Nakamura & Itoh 2008).

Recently, Jill Pruetz and Paco Bertolani reported that chimpanzees at Fongoli, Senegal, used trimmed branches like spears to kill lesser bushbabies concealed in tree holes (Pruetz & Bertolani 2007). The stick described in the article was just like the 'expelling stick' found at Mahale in terms of stoutness, length, and straightness (Nakamura & Itoh 2008). The chimpanzees of Mahale may sometimes kill hyraxes or bushbabies after using such a stout stick to rouse these small mammals from a hole. However, it is still debatable whether or not this type of implement should be called a 'spear'.

Human hunting differs from that of chimpanzees in several ways. Humans actively search for prey targets, kill animals larger in size than themselves, use tools for hunting, and often hunt cooperatively from the outset with the aim of meat-sharing. Chimpanzees typically do not search actively for prey. Except for the probe/skewering stick, they do not use tools for hunting. Chimpanzees begin hunting without assuming that hunted meat will be shared among participants, at least in East Africa.

2.6 SCAVENGING

2.6.1 Riddle of the hominid artefacts at Lake Turkana

Chimpanzees kill and eat living animals. But until the 1980s, as discussed below, no evidence had been presented that they also eat the remains of animals that are already dead.

The study of scavenging behaviour may offer valuable insight into the mystery of human evolution. There is an important archaeological assemblage at Lake Turkana, called the Hippo Site. For a week in September 1971, I stayed at the excavation camp of the University of California, Berkeley, near the lake. This expedition was led by Glynn Isaac. Aside from the hippopotamus fossils, some small scatterings of pebble tools were retrieved, showing that without a doubt the early hominids ate hippopotamus there.

The kinds of tools that were discovered, and their craftsmanship, made it difficult to conclude that they were used to kill

Fig. 2.14 Pebble tools excavated at Lake Turkana site.

hippopotamuses (Fig. 2.14). The archaeologists surmised that hominids used the tools not for hunting expeditions but for butchery (Isaac 1978). We can conclude that they fed on carcasses. And if the early hominids scavenged carcasses, then it is not too far off the mark to speculate that meat-eating apes also scavenge dead bodies. Nevertheless, throughout most of modern primatology, there had been no evidence available indicating that non-human primates were scavengers.

2.6.2 Scavenging chimpanzees

The notion that chimpanzees eat dead bodies came to light in 1980, when Mariko and Toshikazu Hasegawa saw an adolescent chimpanzee carrying the corpse of a blue duiker. After the adolescent chimpanzee discarded the body, they brought it back to our camp for investigation. The corpse was already cold, its eyes clouded, and because there were no wounds they deduced that it was a disease-related death. Out of curiosity, they put the corpse at the provisioning site, and within two hours, 14 chimpanzees had devoured it (Hasegawa *et al.* 1983). Since then, we have observed more than ten other cases at Mahale. I will introduce a few of these cases here.

One day in September 1981, a great pant-hoot chorus occurred early in the morning from the north sector. At this, the entire M-group, which was heading south near the base camp, reversed course and headed north. The chimpanzees in my immediate vicinity frequently sniffed the ground, tossed around dried leaves and twigs, and then, while sniffing, thrust their faces into the underbrush, standing bipedal and inspecting their surroundings. Then they climbed up trees and sniffed even the trunks. This activity went on for a while, before a party of chimpanzees gathered around the foot of a giant tree in the forest. All of the males had climbed this tree and already were divvying up and feasting on a bushbuck (Hasegawa *et al.* 1983).

The chimpanzees had not killed the bushbuck and then taken it up into the tree, so we concluded that the bushbuck carcass was left over from a leopard's meal, and that the chimpanzees had stumbled upon it. The first reason for this inference was that, although not a full-grown adult, the bushbuck was still too big for chimpanzee hunters to handle. Second, the acts of sniffing the scents of the trees and ground suggested that a leopard had dragged the bushbuck's body. Third, the night before we had heard the snarl of a leopard coming from the direction of the meat-feasting scene.

In January 1990, Miya Hamai observed *Kat* and her eight-year-old daughter, *Totzy*, when they encountered the corpse of a blue duiker. The blue duiker was pregnant and it looked as though an eagle had preyed on it. *Kat* removed the foetus and placenta from the duiker's corpse, climbed the nearest tree, and began eating these remains.

But *Totzy*, overwhelmed by the amount of meat, made no attempt to suppress her excitement and began to scream. Not far away were a group of males, so the mother felt uneasy, to say the least. Eventually, the mother panicked, and she and her daughter embraced while letting off a great shriek. Not surprisingly, the adult males approached and seized the prized find that the mother and daughter were holding (Hamai, unpublished observation).

In 1992, when I was tracking a young adult male named *Bembe*, he suddenly climbed a tree, busied himself in the tree's vines for a while, and then clasped a large object. After he descended from the tree, I could see that it was the remains of an adult female red colobus. Although he was holding up the corpse by a leg, the body did not shake or bend in the least, as rigor mortis had already set in. Once on the ground, *Bembe* sped off the path with two associates, disappearing into the underbrush and not making any noise. As some higher-ranking males had been watching from 20–30 m up the path, it was clear that *Bembe* and his associates would have their find snatched away. The corpse was completely unscathed, so the cause of death was apparently disease-related or natural causes (Nishida unpublished observation).

My conclusion is that early hominids likely ate both live and dead animals. The fact that many carnivores, including lion, spotted hyena, and cheetah, are not averse to scavenging on dead animals strengthens our conclusion.

2.6.3 The case of the uneaten corpse

In contrast are those cases in which a discovered corpse is not eaten. On 27 October 1985, we encountered about ten chimpanzees in the underbrush; each was seated facing north and they were not grooming one another. A feeling of tension ran through the group. Eventually, we realised that a large animal's remains lay opposite the brush. It turned out to be a bushpig carcass weighing at least 50 kg. The throat and chest region had been eaten, but the rest of the bushpig was untouched. It was obviously the handiwork of a leopard. It appeared as though the leopard had dropped the bushpig and dashed off, seeing the chimpanzees moving in. The chimpanzees sat in front of the remains for ten minutes or so, but did not lay a finger on the body before getting up and leaving (Nishida unpublished observation).

Letting all that meat spoil would be a waste, so I clasped the tip of the bushpig's hoof so my assistant could make a cut in the hind leg with a panga. What really amazed me was that the splotches on the leg that appeared to be blemishes or freckles were actually ticks (*Ixodes*), which instantly covered my hand. Ticks are always looking for the opportunity to move from a dead host to the next living host. I can assure the reader that the bushpig's leg made a superb pork-chop dinner!

But what on earth were those chimpanzees thinking, with that succulent bushpig right before their eyes? Could it be that they were terrified of the leopard's return? This is unlikely, as it took them over ten minutes to decide to abandon the potential feast.

In August 1999, we saw chimpanzees encounter the fresh carcass of an adult female leopard (see Chapter 4). They did not eat the leopard meat. In August 2005, Kazuhiko Hosaka and colleagues had a good opportunity to observe chimpanzees who found a fresh aardvark carcass (Hosaka *et al.* 2008). This time, no chimpanzee tried to eat the carcass. It appears that leopard, aardvark, and certain kinds of weasel are not included in the dietary repertoire of M-group chimpanzees, and so this carcass also was treated as something to explore.

Regarding meat-eating, there seem to be striking differences between humans and chimpanzees. Humans are hunters who like to eat the meat of big game. In contrast, chimpanzees abandoned a large fresh carcass of bushpig, although they eagerly look for piglets. The chimpanzees of Mahale very often do not eat much from the carcass of an adult colobus monkey, throwing the body away, although they enjoy infant monkeys. On the other hand, the chimpanzees of Taï, Ivory Coast, hunt larger rather than smaller colobus monkeys (Boesch & Boesch-Achermann 2000). However, preference for smaller rather than large colobus monkeys is a definite characteristic of the chimpanzees of both Mahale and Gombe. Why such variations occur remains a mystery to be solved.

2.7 INSECT EATING

2.7.1 Ants: the most important source of animal protein

The chimpanzees of M- and K-groups obtain animal protein mainly from social insects, such as 'acrobatic' ants (*Crematogaster* spp.), carpenter ants (*Camponotus* spp.) (Fig. 2.15), weaver ants (*Oecophylla longinoda*), honey bees (*Apis*), sweat bees (*Trigona* spp.), carpenter bees (*Xylocopa* sp.), and termites (*Pseudacanthotermes*) (Nishida & Uehara 1983).

Crematogaster ants live inside the dead braches of trees (in particular, *Dracaena*). Chimpanzees typically brush away adult ants with one hand, break in two a thick stout branch by grasping each end with a

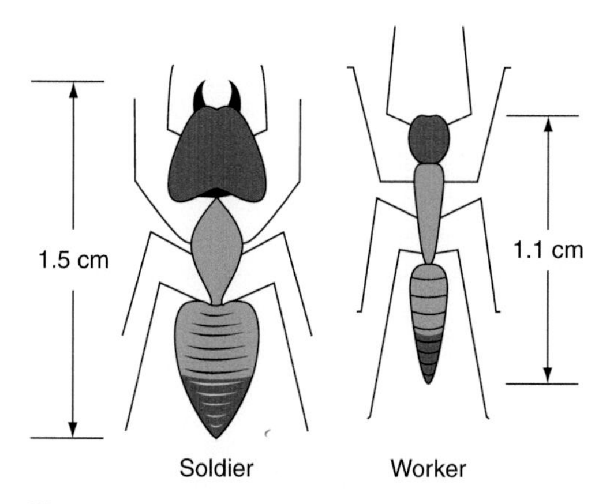

Fig. 2.15 Carpenter ants (*Camponotus*).

hand and pushing the centre with a foot, and lick up the eggs. Although they do not like to eat adult ants, they do not spit them out, even if they are ingested while eating the eggs. Therefore, when sluicing the apes' faeces, we sometimes found a large number of intact ants. The *Crematogaster* ants that adult female chimpanzees eat daily constitute the most important animal protein in their diet.

Carpenter ants are the second most important insect-based food. Chimpanzees use tools for various purposes, and at the top of Mahale's list is fishing for ants (Nishida 1973b; Nishida & Hiraiwa 1982).

Various species of carpenter ant build nests inside tree trunks, and since the entrance to the nest is extremely narrow and deep, it is impossible to pick out ants with the hand or fingers. The chimpanzee thus makes a thin probe and inserts it into the entrance of the nest, where the soldier ants bite onto it. By drawing the probe slowly out of the nest hole, the ants remain 'clamped' onto the probe and snacking begins (Fig. 2.16). This technique is basically the same as the one Jane Goodall discovered being used by the chimpanzees at Gombe to fish for termites.

But ant fishing differs from termite fishing, because the former is usually performed in a tree. Moreover, termite fishing is mostly confined to one period of the rainy season, whereas carpenter ant fishing is a tool behaviour that is performed all year round, making it an act worthy of special mention.

Fig. 2.16 Ant fishing.

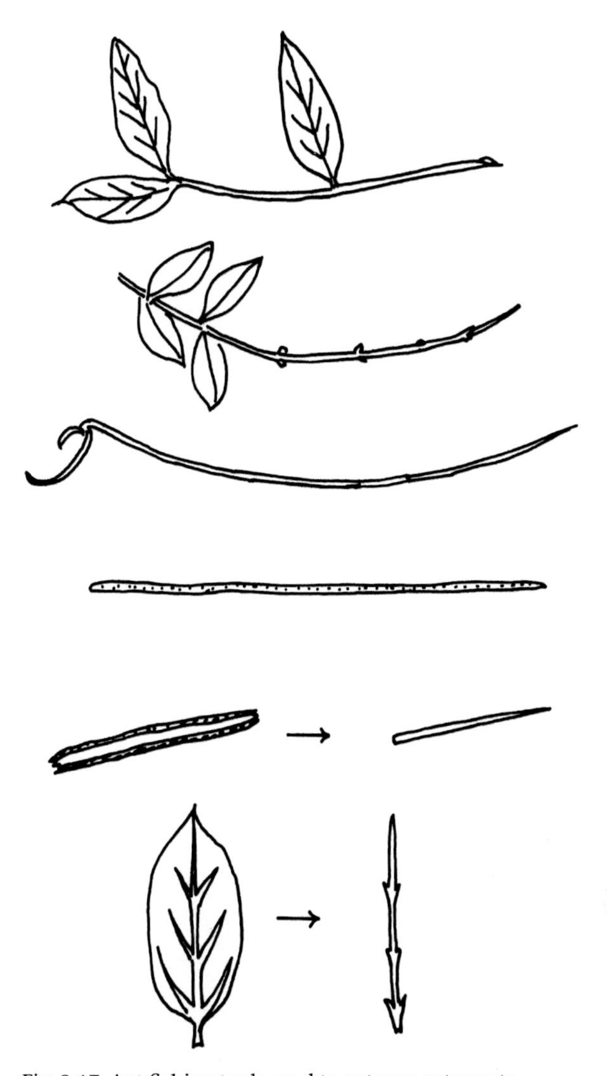

Fig. 2.17 Ant fishing tools used to get carpenter ants.

Various materials are used for probes, including tree branches, blades of grass, vines and vine bark, and the midribs of large-sized leaves (Fig. 2.17). The bark of a specific type of vine is especially widely used, and out of the roughly 100 probes that I gathered, over half of them were made from this special material. If you rip this vine bark vertically, it splits very easily, much like string cheese, and because it is highly flexible, it is popular among the chimpanzees.

Weaver ants make their nest in a tree by affixing green leaves together with the sticky silk-like substance produced by larvae. Since weaver ants do not sting, but instead bite ferociously, chimpanzees climb the tree, grab the whole nest, descend from the tree quickly, open the nest and eat the ants in haste on the ground. While devouring the ants, they constantly use one arm to brush away the ants biting their hands, arms, and elsewhere. The Kitongwe name for the weaver ants, 'sitetambolo' (bite penis), suggests that these ants sometimes bite even humans in sensitive areas!

2.7.2 Local differences in ant eating

The chimpanzees of Mahale eat carpenter ants. They not only fish for the species that nest inside tree trunks by inserting a probe into the nest, but also bite open grass stalks to eat the species that nest inside dry grasses. Although there are carpenter ants at Gombe, the chimpanzees there typically do not eat them. The eggs, larvae, and pupae of *Crematogaster* ants can be found in the dead branches of trees such as *Dracaena*, and these are eaten everyday, putting them at the top of the insect menu (Fig. 2.18). However, it is said that the chimpanzees of Gombe almost never have been seen to eat *Crematogaster* ants. On the

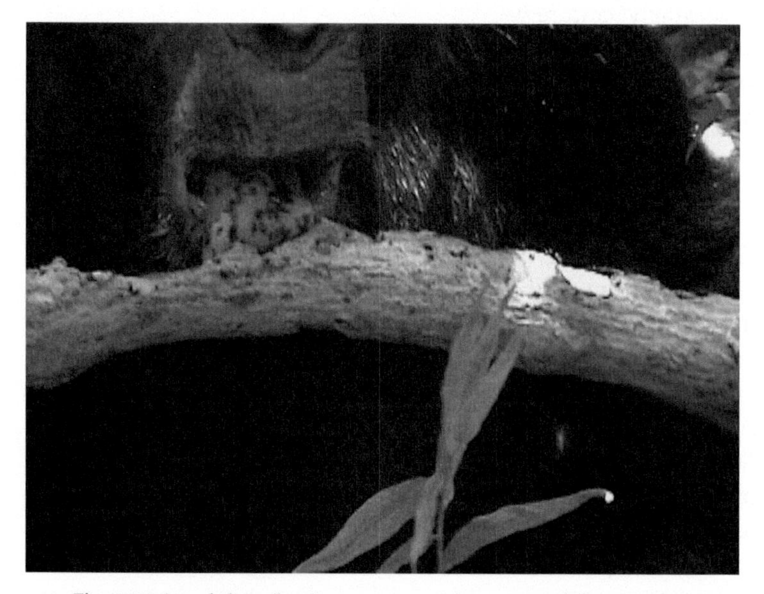

Fig. 2.18 An adult male, *Carter*, gnaws a dry, stout stick to eat the eggs of *Crematogaster* ants.

other hand, if the Gombe chimpanzees see a procession of driver ants (*Dorylus molestus*), they use a sturdy 1 m wand to 'stir up' the ants in their assembly. The soldier ants become angry and climb up the wand. When the front-line of the defensive horde gets near the hand holding the wand, the chimpanzee lifts up the wand and with the other hand sweeps the ants into a ball and stuffs them in his mouth (McGrew 1974). These soldier ants are a ferocious lot, and they do not let go once they clamp onto you.

However, when the chimpanzees of Mahale see a mass of driver ants, they take every precaution not to step on them, and they never try to eat them. Why do such differences occur?

I hypothesise that driver and carpenter ants are nutritionally comparable, and that at Gombe there may be more driver ants than carpenter ants; consequently, the chimpanzees there have more chances to experiment with dipping for driver ants than with fishing for carpenter ants. At Mahale, the different conditions suggest that there are more carpenter ants, and so there are more chances to experiment with fishing for them. When the dipping technique at Gombe or the fishing technique at Mahale was established, the chimpanzees of either site may have lost interest in developing the skills needed to gather the other type of ant, as they had acquired the ability to obtain sufficient nutrition from their preferred type of ant prey. Accordingly, culture can be influenced by subtle environmental variations, but what is important is that once one pattern is selected by a group, it becomes standardised as a tradition accepted by the group's majority (Nishida 1987).

2.7.3 Termites

The termite fishing often observed at Gombe and other sites is not observed within the home ranges of K- and M-groups of Kasoje forest. In fact, the only person who ever saw these groups fishing for termites was Shigeo Uehara (1982). The fished and eaten termites observed by Uehara were not of the genus *Macrotermes*, but of the genus *Pseudacanthotermes*, while the termites fished by chimpanzees at Gombe are *Macrotermes*. But B-group, found to the north of K-group's range, fishes for *Macrotermes* termites just at the start of the rainy season (Nishida & Uehara 1980; McGrew & Collins 1985). In K- and M-groups' ranges to the south of B-group, however, one finds the genus *Odontotermes*, rather than *Macrotermes*. It appears that *Macrotermes* and *Odontotermes* occupy similar competitive ecological niches.

Odontotermes builds similar but much lower towers, but the lifestyle of *Odontotermes patruus* differs from that of *Macrotermes*, and, like some other members of the genus, it may produce a more distasteful defensive secretion than does either *Macrotermes* or *Pseudacanthotermes*.

Pseudacanthotermes spiniger is a species that is captured easily without fishing tools. K-group chimpanzees regularly catch the winged reproductive form of this species in the latter part of the rainy season (March and April) by simply toppling the termite mound's towers by hand. The differences in the termite fauna, particularly the structure of their towers, probably explain why the most intensively studied chimpanzees of K-group have been observed only rarely to use tools to obtain termites.

Why do B-group chimpanzees use tools to exploit termites, while neighbouring chimpanzees of K- and M-groups do not? The difference seems to arise from the local heterogeneity of the home ranges of the neighbouring unit-groups. K- and M-groups' ranges are on the western side of the highest peaks of the Mahale Mountains and enjoy high annual rainfall, with a mean of roughly 1800 mm, producing thicker vegetation. B-group's range is on the northwestern edge of the lower peaks (as the highest elevation is about 1600 m), and its annual rainfall is only about 1350 mm, resulting in much drier vegetation types. This environmental difference most certainly influences the termite fauna of the two ranges.

In 1977, Takayoshi Kano invited me to visit his study site of pygmy chimpanzees at Wamba. What fascinates me is that the people of (then) Zaire gathered termites of the genus *Macrotermes* in a very similar fashion to the chimpanzees. However, there were differences. They used a machete to widen the entrance of the termite nests. They also brought hot charcoal to push hot air deep into the nest chambers. Then, they used a duster-like fishing tool made of bark, rather than a simple probe, to fish simultaneously for many termites (Fig. 2.19). Thus, they used three kinds of tools, or a toolset: machete, fire, and duster. Moreover, they did not eat termites on the spot, but collected a large number and took them home to cook in hot water.

Recently, however, it has been shown that chimpanzees also use a toolset. Shigeru Suzuki found evidence that when the chimpanzees of Ndoki, Republic of Congo, seek to feast from a termite mound, they use a heavy-duty probe just shy of a metre in length, jab it in, make a hole, and then insert a separate flimsy probe to manoeuvre inside the hole (Suzuki *et al.* 1995; see also Sugiyama 1995 for toolsets in different contexts, called 'tool composites'). His point is that to

Fig. 2.19 Mongo women fishing for termites.

fulfil a single goal, these chimpanzees utilise two different tools in succession. More recently, in the Goualougo Triangle, Republic of Congo, Crickette Sanz and David Morgan installed 20 video-cameras at various places within a range of a group of chimpanzees, and found that from the outset, chimpanzees appeared at termite mounds with two kinds of tools, thick and thin sticks, in their mouths (Sanz *et al.* 2004). They dug with the stout stick to monitor and penetrate into termite tunnels and then threaded the probe-like thin stick into the hole to fish for soldier termites. Thus, Sanz and Morgan not only confirmed Suzuki's reconstruction of behaviour but also discovered the use of a foot, in addition to the two hands, when pushing a stout stick into the ground, in order to perforate a subterranean nest of termites. Seeing their video, I was struck by its similarity to the hoeing of human farmers (McGrew 1983, 1992)!

2.7.4 Other insects

The above species do not represent all of the insect diet eaten by the chimpanzees of M- and K-groups. First, attention must be given to the larvae of hemipterans (*Phytolima lata*). The larvae of these insects infest the leaves, buds, flowers, and fruits of *Milicia (Chlorophora) excelsa* and make small galls, which chimpanzees spend more than two hours per day eating in August and September every year. The same habit was first recorded at Gombe (Wrangham 1975).

Insects that are occasionally eaten include the larvae of coleopterans, imagoes of orthopterans, and ants of the genera *Tetramorium* and *Monamorium*. It is likely that chimpanzees consume fig wasps (*Blastophaga* spp.) when eating figs of various species, even if by accident. Chimpanzees sometimes poke a stick probe into a hole of a dry branch. Many times, adult carpenter bees thus roused fly out of the hole. Therefore, it is natural to conclude that the apes sometimes eat the larvae of these bees. Local differences in insect food and harvesting techniques have been described in greater detail by McGrew (1983, 1992).

2.8 DAILY ACTIVITY RHYTHM

We usually have three meals: breakfast, lunch, and supper. However, as far as I could determine in October 1965, when I arrived at Kasoje for the first time, the village people had only one major meal (of *ugali*, or hot manioc porridge) between 11:00 a.m. and 1:00 p.m. Of course, they occasionally had snacks such as raw sweet manioc, bananas, or some fruit. I needed about 40 workers for various tasks, including the clearing of bush to make space for a sugarcane plantation. In the process of negotiating working hours and wages, they asserted strongly that they should begin work at 7:00 a.m. and finish around noon, although I asked them to begin at 8:00 a.m. and end at 3:00 p.m. This made me realise that they had no custom of taking breakfast and therefore could not work until much later than noon. Finally, we agreed that working hours would begin at 7:00 a.m. and end at 1:00 p.m.

How do chimpanzees allocate daily hours to various kinds of activities? They rise at dawn between 6:00 and 6:30 a.m., leave their night beds, and excrete from a branch a few metres away. *Ndilo*, a mother with an infant, *Nick*, however, once got up at 8:10 a.m., which is the latest waking time I have recorded.

After waking up, chimpanzees often have a brief period of grooming, or they may climb down the tree and begin long-distance travel to a feeding place or begin to feed near the bed-site. Intensive feeding continues on and off for three hours or so. Before the hot midday hours, chimpanzees take a rest on the ground in the dry season and often in trees in the rainy season. This is the time when they groom one another or get some sleep. Youngsters play at wrestling and chasing, or climb, hang-wrestle, and make their playmates fall. Smaller infants climb, brachiate, and leap down or fall to the ground one after another.

Around 3:00 p.m., chimpanzees wake up, pant-hoot, and begin to travel a long distance to the next food patch. When they arrive there, they have a late afternoon feeding period. This lasts for more than three hours, and then they often move to a forest of *Pycnanthus*, if available, climb a tree, make night beds, emit bedding pant-hoots, and sleep. This is a typical daily rhythm, featuring two intensive feeding periods with one long intermission in between, when food is abundant and the weather is calm. When the sugarcane was abundant in the newly made plantation in May–August 1966, K-group's chimpanzees used to visit there twice a day, in the morning and late afternoon (Nishida 1972c). They rested in the forest close to the plantation during the hot midday period (Fig. 2.20). Consequently, at that time I obtained a very clear daily feeding schedule for the K-group chimpanzees.

However, when I began to follow female chimpanzees on an individual basis, I found that much variation occurred from day to

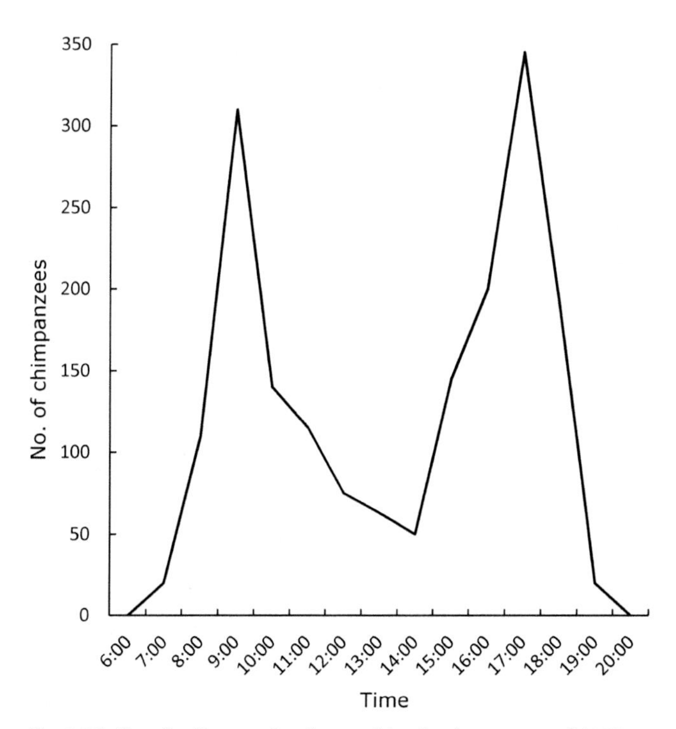

Fig. 2.20 Two feeding peaks observed in the dry season of 1966 (adapted from Nishida 1973b).

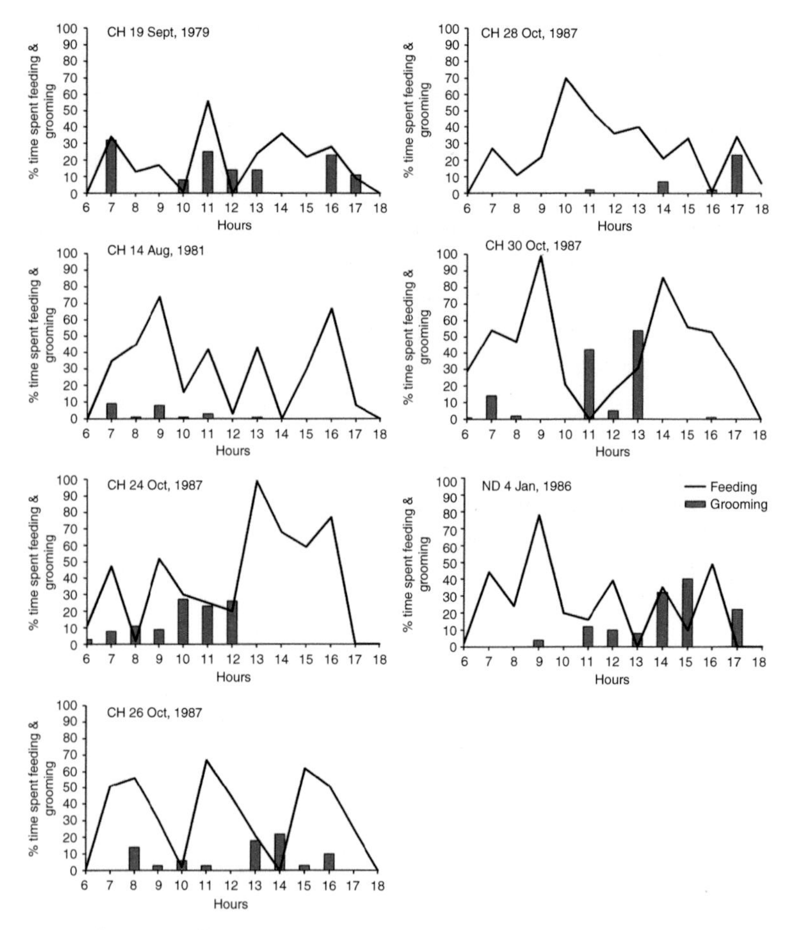

Fig. 2.21 Daily activity rhythms shown by two adult females, *Chausiku* and *Ndilo*.

day, in particular, when food was scarce, namely during the dispersing season. I followed two K-group-born females, *Chausiku* and *Ndilo*, from bed to bed as targets six times and one time, respectively. If an individual spent more than 50 per cent of the time feeding in any one-hour period, I operationally defined this as a major feeding hour (M), versus more than 30 per cent but less than 50 per cent as a minor feeding hour (m). *Chausiku* showed five patterns over six days: 1M 4m, 1M 2m, 1M 2m, 2M, 2M 2m, and 3M. *Ndilo* showed a 1M 4m pattern (Fig. 2.21). Thus, they had 2–5 intensive feeding periods depending on the day, showing no clear-cut pattern in the daily activity rhythm. Obviously, heavy rain had a considerable influence on their activities, causing them almost to stop eating, although light rain apparently had no such effect. In addition, the presence or absence of dependent offspring or group-mates and the quality and distribution of food may influence the daily activity rhythm.

Humans living in developed countries carry out working lives that require us to take two or three regular meals a day. However, for chimpanzees, eating food is their major activity, and there is no reason to eat on a strict schedule. What should be taken into consideration are feeding competition among animals of the same and different species and avoidance of activity during the hottest hours of a day. The early morning meal is most important in terms of hunger and competition for ripe fruits. The evening meal is necessary to store up energy for the 12 hours of fasting during the night.

I propose that humans employed a two-meal system only after the invention of agriculture in settlements, as chimpanzees do in the season of food abundance. Then, together with the beginning of civilisation accompanied by much greater food abundance, some civilised peoples living in cities began to add lunch as a luxury. 'Brunch' is not new, because it shows only a return to the old two-meal system!

2.9 PREDATION ON CHIMPANZEES

2.9.1 Lion eats chimpanzees

For a long while, it was believed that aside from humans, chimpanzees had no natural enemies. It is not absolutely inconceivable for an infant chimpanzee to fall victim to a leopard or eagle, but no proof existed that any other animal besides humans ever ate chimpanzees. So far, I have not seen any predator eat a chimpanzee. Regarding other non-human primates, I have only once seen a crowned hawk-eagle eating a red-tailed monkey.

In 1966 Junichiro Itani and Takayoshi Kano observed a group of chimpanzees in the Ugalla region that were wailing in a treetop, probably after being chased there by the lion that Kano inadvertently collided with, head-on (Itani 1979; Kano 1972)!

Takahiro Tsukahara, a graduate student in the Department of Anthropology of the University of Tokyo, furnished proof of such predation, removing any doubt. After two lions were spotted in the vicinity of the Kansyana camp in 1984, the following five years were quiet. At Kasoje, 1989 came in with an ominous roar. In March, Hitoshige Hayaki reported hearing a lion's roar in the south sector of M-group's range.

After that, Tsukahara, who was on his own in Kasoje, worried that a lion had begun dwelling there. The damage done is described here (Tsukahara 1992).

First, the chimpanzees began to use the hilly section of the group's territory, rather than their usual ranging routes, which made human tracking a burden. Second, the cook and research assistant were terrified of the lion, and so they stopped coming to work early in the morning, as lions are most active then. From the village to camp was about a 1 km trudge through the bush, which these employees now would not take. To Tsukahara, trekking alone in the hills meant grave danger, but waiting on the Tongwe workers meant the research work could not start until noon. This dilemma caused him to lay aside the core of his research for the time being (which was the grooming relationships among adult male chimpanzees), and search for a solution as he investigated the lion's next move. He formed a patrol regiment with one of the assistants (who carried a rifle), which included poking around here and there, implementing a painstaking documentation system citing locations of faeces, vocalisations (roars), and paw-prints, plus the number of lions, the direction of their movements, and their home range.

The only way that black hairs were discovered in the lion's faeces was as a result of the unplanned exertions made by Tsukahara. His intuition told him that those long black hairs were those of a chimpanzee. I ordered him to send me, by airmail, a portion of the hair. As there are other animals at Mahale with black hair, such as blue monkeys and bushpigs, one could not conclude that just because the hair was black it must have belonged to a chimpanzee. I sent the hair samples off to a hair specialist, Haruhisa Inagaki. From his former work at the Japan Monkey Centre, he had accumulated samples of hair from captive animals. He matched the hairs up under an electron microscope and informed me that they were unmistakably from a member of Hominoidea. An electron microscope can discern hominoid (ape and human) hair from cercopithecoid (Old World monkey) from the medulla structure of hair (Inagaki & Tsukahara 1993). Since no human had been killed by a lion around Mahale, the hair must have come from chimpanzees.

That was an exciting bit of news, but not the kind of finding that pleased me on a personal level. Tsukahara was good enough to keep me posted on the movements of some of the chimpanzees to whom I had grown particularly attached: 'Recently, *Wasobongo* is nowhere to be found' or 'I have not seen *Mtwale* these days.' As it was not unheard of that some chimpanzees, in particular those females that keep to M-group's southern sector as their core area, were not seen over a three-month period, I at first did not worry too much about the news. Later, I became gravely concerned because, although I wished these to be cases of temporary absence, it was beginning to look like they were the victims of lions.

2.9.2 At least eight taken as victims

Tsukahara suspected that the last days the chimpanzees were seen and the days that he discovered black hairs or bone in the lion faecal samples would fit the same timeline. After returning to Japan, he paid a visit to the Ueno Zoo in Tokyo (Japan's largest zoo) and fed one of the lions there in order to investigate the length of time it takes for food to pass through the gut after it has been eaten. He found that the shortest duration was 24 hours, while the longest was 60 hours. He also asked Tokyo University morphologist Gen Suwa to run an analysis of the teeth and bone that were discovered with the hair. The teeth and bone samples proved to be stronger evidence than were the hair samples, as Suwa was able to determine the ages of two of the chimpanzees that were eaten from these samples.

Thus, two mother-and-child pairs and two adolescent males, at least six individuals in total, were confirmed as victims of the lions. In the following year, 1990, two more victims were taken. Over two years, I learned of the deaths of three female chimpanzees with whom I had developed relationships over 20 years: *Chausiku*, *Ndilo*, and *Wantendele*. There was also another female victim I had been close to for 17 years, *Wasobongo*. The motivation and power to go on conducting research on female chimpanzees, as I had done since 1979, no longer resided within me. I forever stopped systematic study of adult female chimpanzees.

It is possible to interpret the significance of these incidents in one of two ways. One consideration is that the lion is an ancient predator of the chimpanzee. The other is that this was a fortuitous accident – lions seldom attack humans, but at times, just as a particular man-eating lion emerges (Patterson 1979), so too can a chimpanzee-eating lion. For now, we cannot draw a conclusion about either, though I believe the former is the case. That is, observations made in savanna woodland such as Kasakati or Filabanga, where there are many lions, show that chimpanzees gathered and dispersed freely within riverine forests, but when moving from one riverine forest to the next, traversing the savanna's open canopy forest, they travelled in large parties

(Itani & Suzuki 1967; Izawa 1970; Nishida, unpublished observation). Is this arrangement done for anti-predator purposes, in particular against lions? For a long time, as lions were nowhere to be seen around Kasoje, the chimpanzees did not take the appropriate anti-predator counter-measures, highly increasing the possibility of numerous casualties. Almost every behavioural pattern of a chimpanzee is modified by experience after it is born (Chapter 9). Therefore, although the avoidance of lions is innately motivated, chimpanzees need traditions when it comes to coping appropriately with formidable predators. Perhaps the chimpanzees of Kasoje lost this tradition long ago because they rarely met with lions.

Tsukahara's discovery was a landmark in that it clarified the point that when it comes to the social structure of great apes, we cannot ignore the effect of predators.

Let us return to Kasoje. Tsukahara encountered lions three times in the bush. In addition, some of the video crew from Japan also had an encounter with an adolescent male lion early in the morning. All of the assistants in the camera crew were so startled that they fled – after throwing away all of the equipment! The following year Kenji Kawanaka met up with a young male lion in the early morning hours on his way to the mountains and he ran for his life, forgetting the Tongwe proverb that said 'Never show your buttocks to a lion' (that is, do not flee to escape from a lion). Fortunately, the young lion seemed to be satiated, and he emitted only a slight bark.

When I arrived at Kasoje in October 1989, the lion was still on the prowl. At about 4:00 a.m. on 5 November, a lion's roar could be heard a few hundred metres west of the camp. After dinner, from about 8:00 p.m. until after 11:00 p.m., close to ten roars could be heard. Chimpanzees who made beds in the Kansyana Valley responded to the roars with what I call the 'fear call', a wispy vocalisation, but after several roars this changed to pant-hoots, and finally, even though they could still hear roars, they stopped replying. Luckily, I never ran into the lion. From 1991 onward, the lion never again showed up in Kasoje.

2.9.3 Chimp–leopard relations

Around the same time, at Taï, Ivory Coast, Christophe Boesch observed that several adult chimpanzees were wounded and some were killed or injured by leopards (Boesch 1991a). Thus, a feline predator's importance to chimpanzee social structure was confirmed in both East and West Africa. What really puzzles me is this: why are the bigger chimpanzees of West Africa preyed upon by leopards while their smaller counterparts in East Africa are not? At least in the leopard faeces that we have collected at Mahale, we have not found the black hairs of chimpanzees.[3] It would be worthwhile to investigate whether the leopards of West Africa are larger than those of East Africa.

It is certain that Mahale chimpanzees regard the leopard as an enemy. As we will see in Chapter 4, there have been observations of Mahale chimpanzees even abusing the carcass of a female leopard.

[3] In addition to me, Nobuyuki Kutsukake and Fumio Fukuda collected leopard faeces, but found no evidence that chimpanzees were eaten (unpublished information).

3

Growth and development

3.1 DEVELOPMENTAL STAGES

The developmental stages of a chimpanzee's growth are divided into four phases and can be defined as: infancy, juvenility, adolescence, and adulthood (Hiraiwa-Hasegawa *et al.* 1984; Goodall 1986; Nishida *et al.* 1990).

Infancy (0–4 years old) entails being carried around by the mother, breastfeeding, being hugged when it is cold or raining heavily, and sharing an overnight bed.

Juvenility (5–8 years of age) comes after weaning is complete. The juvenile is physically independent, walks and climbs on its own, and makes and sleeps in its own night bed, but stays in close association with its mother. A juvenile is still under its mother's supervision, especially when it comes to ranging.

Adolescence is marked by sexual maturity such as dramatic enlargement of the testes in males and tumescence of the sexual skin in females, but adolescents lack full adult body-size. Adolescent males are 9–15 years old, and females are 9–12 years old. They are still socially immature, and usually are not permitted to join in adult grooming clusters.

Adulthood is when individuals are socially and physically mature. Males are 16 years old and over, and females are 13 years old and over. However, development is a progressive phenomenon, and as each individual's pace differs, the age classification above is only a broad outline.

3.2 INFANCY

3.2.1 Newborns

During the perinatal term, many mothers separate from other members of the group and largely spend time alone. Mariko Hasegawa, who

Fig. 3.1 A newborn baby.

studied the mother–infant relationship, called this period of isolation 'maternity leave', which sounds familiar to us humans. When the mother first returns to the group after giving birth, she keeps space between herself and the adult males, greeting them repeatedly with pant-grunts in an anxious manner. The males lower their heads in order to size up the new addition to the group, and sometimes make charging displays. An alpha female, *Wakampompo*, did not take maternity leave, but continued to associate with adult males, including the alpha male, *Ntologi*, during pregnancy. I found that *Wakampompo* often adopted a prone posture with all arms and legs flexed on the ground during the perinatal period.

A newborn baby chimpanzee is very small. The upper half of the face is blackish and the lower half of the face is paler, like the palms of the hands contrasted with the reddish feet (Fig. 3.1). Hair on the head and body is thin, and as the baby is nearly naked, its whitish skin is visible. Many people imagine that chimpanzee skin is black and so our ancestor's skin was black, implying that African people have 'primitive' features. This is quite the opposite. Human ancestors likely acquired black skin as an adaptation to a high ultra-violet (UV), sunny savanna environment. When human ancestors invaded temperate regions, they re-evolved the older characters of white skin colour in order to absorb UV rays.

Immediately after birth, the baby moves its head in search of the mother's teat, a behavioural pattern called 'rooting'. During this

Fig. 3.2 A mother, *Chausiku*, transports her newborn son on her head.

period, suckling is irregular, and the baby constantly lets out whimpers and staccato vocalisations, voicing frustration if not on the nipple. For a few weeks after birth, the baby's motor functions are minimal, as are its abilities to grasp the mother. Its leg control is especially under-developed, and when the mother is walking on the ground, she presses her baby to her belly, with one hand supporting its back; in the treetops she crosses both legs, keeping her baby between them while brachiating. On a downgrade, she keeps the baby between her thighs, bracing herself with both fists clenched to the ground, while she moves her torso. This mode of transport is called 'crutching', because she uses both arms as if they were crutches.

The mother chimpanzee holds her baby against her belly and carries him, but *Chausiku* wore her one-month old on her head like a beret (Fig. 3.2). If this birth had been her first, then such bizarre transport would be understandable, but this infant was her third!

3.2.2 Tranquillity of the mother–child relation

A baby chimpanzee usually begins riding its mother's back after the first three months, but always within the first six months of life, and except for times of danger, dorsal riding is typical by its first birthday. A mother chimpanzee whose baby is on her belly may either flip it over her head and put it on her back with her hands, or will shove it towards her back with one hand. Over time, the baby comes to enjoy riding on its mother's back. From this position, it is much easier for it to take in the world around it, plus it never has to worry about bumping its head on the ground below. When the infant is carried ventrally, it clings to the mother's belly, grasping her sides with both hands.

At six months of age, feeding frequency starts to stabilise, with a typical suckling rate being about 1.5–2 minutes per hour, with a suckling bout from each nipple. This continues until three years of age. Interestingly, infants not only tend to suck from the left nipple more often and for a longer time, but also they begin a suckling bout from it rather than from the right nipple. Mothers seem not to lead the infants to the left nipple, but rather infants themselves prefer the left nipple (Nishida 1993a).

A mother may confiscate a plant-part not in the group's food repertoire from her infant. I call this 'education by discouragement' (Nishida 1987). However, during later studies, I did not routinely observe such maternal interference. So, 'education by discouragement' is not systematic and perhaps depends on the mother's unpredictable character. I never saw 'education by encouragement'. Although Christophe Boesch described a mother chimpanzee's actions in relation to her daughter's hammering of nuts as 'active teaching', he recorded this only once (Boesch 1991b). Chimpanzees lack obvious teaching as a systematic method of learning.

A small infant seems to learn lots of social attitudes from its mother. For example, a newborn infant utters staccato calls from the mother's belly when she pant-grunts. Within the first year of life, an infant on the mother's back pant-grunts to an adult male to whom the mother is pant-grunting. In this process, an infant probably learns whom and how to greet others. By the third year of life, an infant pant-grunts to adult males even before the mother does.

Until about two years of age, an infant chimpanzee's most usual playmate is its mother. She sits with her infant in her lap and pokes its chest or belly, tickling it. One morning I heard playful panting, 'Ha, ha, ha, ha', but could not see the figure of a chimpanzee anywhere. I finally realised that the laughter was coming from a bed overhead. By climbing up a tilting tree and peeking in, I saw a young mother, *Plum*, sprawled out on her back, tickling her first infant who was yelping with laughter – that is, play-panting.

After two years of age, an infant begins to sit up straight on its mother's back, stand on all four limbs, or lie on its side, trying

Fig. 3.3 A two-year-old rides in typical dorsal position.

Fig. 3.4 A two-year-old stands on all fours on the mother's back.

various positions (Figs. 3.3 and 3.4). This is also the time when an infant begins walking on its own, either in front of or behind the mother in long-distance travel, although it often returns to her back. The infant becomes much more active, playing with its peers in the treetops, and even when the mother calls it to come down, it may not do so. When this happens, the mother makes soft sounds, 'Hoo hoo hoo', in the

Fig. 3.5 The mother drags her infant with one hand.

infant's direction, beckoning it to descend at once from the tree; if the infant still does not come down, she climbs the tree and drags it down (Fig. 3.5).

Once a mother chimpanzee, *Fatuma*, repeatedly dragged her daughter *Penelope* out of a tree, but every time *Fatuma* put her up on her back, *Penelope* kept escaping. Getting fed up with this, *Fatuma* grabbed *Penelope* by the hand and dragged her for over 5 m, holding her daughter with her upper arm to her belly and carrying her off in a wrestler's grasp.

Infants over two years old like energetic play. Their mothers chase them for a bit and then after nabbing them, tickle them robustly with their mouths. A favourite game of infant chimpanzees is to leap from their mother's backs while in motion, speed off ahead, climb a fallen log, wait for mother's arrival and then . . . *pounce*! They jump onto the mother's back and repeat the sequence over and over.

From three years of age, excluding long-distance walking, a mother carries her infant less and less. When the group is travelling, three-year-olds do somersaults while on the move. They occasionally also make pirouettes. They stay within the mother's sight, but when immersed in play an infant may go for two hours at a time without breastfeeding. Until this time, grooming was always one-way, done by the mother, but now the infant occasionally grooms its mother for a few minutes at a time. Mothers may wrestle whole-heartedly with their older infants (Chapter 4). During play, a mother may also grip the infant's hand between her teeth and drag it.

This peaceful mother–child relationship abruptly stops at four years of age, when the mother resumes oestrus and the infant approaches the weaning stage.

3.3 DEVELOPMENT OF FEEDING TECHNIQUES

3.3.1 Gathering techniques for plant food

In the first three months of life, infants depend solely on their mother's milk. An infant of four months was once seen to put into its mouth a pith of a ginger stalk chewed and abandoned by its mother; an infant of five months put a *Garcinia* fruit into its mouth; and another infant of five months put an unripe fig into its mouth. However, in none of these cases was food ingested. Also, infants of five and seven months licked the dry wood of *Pycnanthus* or *Garcinia*. At one year old, infants begin to ingest fruit, resin, blossom and ginger pith, as well as licking dry wood. Two-year-old infants skilfully consume leaves, holding a branch with one hand while pulling off leaves with the other. They try to eat everything that their mothers are eating, from fruits and leaves to termite soil, ants, and colobus meat. Three-year-old infants ingest almost anything, but some foods cannot be processed easily. Some of these hard-to-process foods include compound fruits such as *Myrianthus*, thick/hard-shelled fruits like *Voacanga*, *Saba*, and *Strychnos*, or the pith of tough herbs such as *Marantochloa*. Moreover, they are not yet skilled in eating petioles of *Ipomoea*, an important plant in swampy habitats. They still frequently recover food remnants abandoned by their mothers. Four-year-olds can eat *Saba* fruit, the most important food of M-group chimpanzees. Thus, by this age, their feeding techniques of plant foods are almost complete.

3.3.2 Techniques for eating insects

When eating *Crematogaster* ants, chimpanzees select a dry, stout branch that has lots of pupae and eggs. They obtain the bough from the bush or in vine tangles above the ground, then bring it to the ground. They first hold it with one hand and rub its surface with the other in order to brush off attacking adult ants. Then they hold each end of the dead branch with a hand and push a foot to the middle of the log in order to break it open. Youngsters less than three years of age typically cannot do this. Only one three-year-old female has been seen to succeed, but some four-year-old infants have mastered this log-breaking technique.

Weaver ants (*Oecophylla*) seem to be the most difficult insect to eat. Ants are contained in a nest of woven leaves a few metres above the ground. A chimpanzee plucks the nest, quickly climbs down to the ground and repeatedly puts the nest to its mouth. Ants aggressively bite the chimpanzee. Only adult or adolescent chimpanzees can resist the ants' painful attacks. Juvenile and infants approach only destroyed nests and pick off the few weaver ants that remain.

3.3.3 Technique of fishing for carpenter ants

An infant chimpanzee of Mahale, in the first years of life, seems to pay little attention to its mother's ant fishing behaviour. A one-year-old infant takes interest in a fishing probe prepared by the mother or siblings, but only holds it between its lips, doing nothing further. A two-year-old infant begins to show many ant fishing behavioural patterns for the first time: it peers at others (especially the mother or sibling) while they fish for ants, grasps ants with the palm and eats ants directly by mouth or by mopping with the back of the hand. An infant may try to snatch a fishing probe being used by its mother or sibling, thus disrupting the fishing activity, or hold a fishing probe in its hand. It may try to poke fingers into the ants' nests. Three-year-old infants begin to push a fishing probe into an ants' nest and then pull the probe from it. They acquire fishing probes from close relatives or take abandoned probes left at the ants' nest entrance. They modify probes by clipping them with their incisors. Their fishing probes are typically short. Only a few three-year-olds succeed in fishing for ants, and even they do so rarely.

Many four-year-old and most five-year-old infants successfully fish for carpenter ants. Four-year-olds make fishing probes by removing leaves or tiny branches, or use a leaf-clipping technique. *Rubicon*, a female with deformed fingers, fished for ants by the age of five years. Thus, most chimpanzees have mastered the basic ant fishing technique by the time of weaning and can process many difficult foods.

However, five-year-olds still abandon ant fishing when bitten by many soldier ants, suggesting that their defence against soldier ants remains incomplete at this age. The skill of fishing is mastered by the age of seven, but a seven-year-old male, *Oscar*, was an exception, as he was never seen to fish for ants.

3.4 FOOD SHARING BETWEEN MOTHER AND OFFSPRING

3.4.1 Mastering the food repertoire

Food sharing is relatively rare among non-human primates, but not so if we consider the whole animal kingdom. Bird parents feed fledglings, and parents of carnivores such as cheetahs and hunting dogs feed their cubs (Dugatkin 1997). In primates, many monkey species of the marmoset and tamarin family (Callitrichidae), such as lion tamarins, provision their infants with insects, such as grasshoppers (Goldizen 1987). White-faced capuchin mothers occasionally allow their infants to eat from their food (Perry 2008). Besides parents, 'helpers' may also feed infants or fledglings; most of the time the helpers are older siblings (Goldizen 1987).

I will discuss food sharing among adults in Chapter 8. Although food sharing occurs among adults, mother and infant food sharing occurs most often in chimpanzees.

Food sharing occurs only when the infant chimpanzee begs for food. A mother never spontaneously gives something to eat to her infant. This pattern differs greatly from a human mother's sharing. The baby chimpanzee begs for food by putting its mouth close to the mother's, reaching for her food, or by putting its hands to the mother's mouth (Silk 1978; Hiraiwa-Hasegawa 1990). This gesture is usually accompanied by a plaintive cry called 'whimpering'.

Depending on the mother, food sharing usually begins when an infant is six months old, and the rate of infant solicitation for food dramatically increases in the second year of life, then gradually decreases, virtually disappearing by the seventh year (Fig. 3.6). The ontogeny of food sharing precisely follows that of solicitation, because mothers share food only when requested to do so by their infants. However, the success rate of food solicitation (sharing frequency/solicitation frequency) does not change markedly across ages (Fig. 3.7). The probability that a mother will comply with her infant's begging slowly goes up as the infant matures; between ages four and five years, it is almost 100 per cent. This seems inconsistent with the food sharing frequency curve described above, but it is not. As the infant grows older, it is not that the mother denies it food, but rather that the frequency of begging declines. As a result, food sharing frequency slumps. Sharing success rate remains high because weanlings often beg but only for rare and highly attractive foods like meat, and the mother inevitably gives in to her infant's violent tantrums.

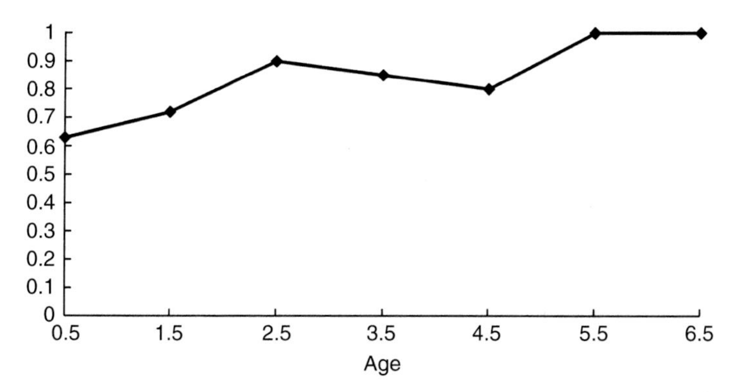

Fig. 3.6 Frequency of food sharing by mother as a function of infant's age in years. Number of solicitation bouts and sharing bouts per number of feeding bouts. Age categories: 0.5 = 0–1 year old; 1.5 = 1–2 years old; 2.5 = 2–3 years old; 3.5 = 3–4 years old; 4.5 = 4–5 years old; 5.5 = 5–6 years old; 6.5 = 6–7 years old (from Nishida & Turner 1996).

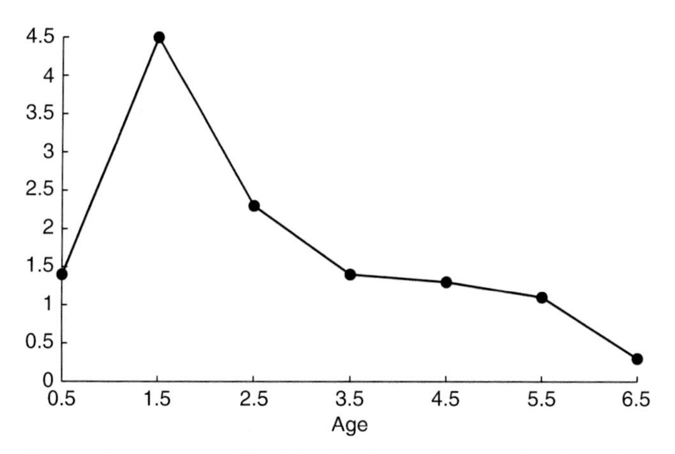

Fig. 3.7 Success rate of food sharing by mother as a function of infant's age in years. Number of sharing bouts per number of solicitation bouts (from Nishida & Turner 1996).

Three- or four-year-old infants rarely beg for food, because they can obtain almost all food themselves. If these infants do beg for food and mothers do not share, they protest, roll on the ground, and throw a tantrum. The target of such a request is almost always meat or a few other precious foods that infants and juveniles can never obtain. These tactics have a nearly 100 per cent success rate in food solicitation.

Food can be divided into 'difficult food' and 'easy food'. Shared food that infants eat is typically food that is difficult to obtain or food that needs to be processed before being eaten (Silk 1978; Hiraiwa-Hasegawa 1990). 'Difficult food' includes: meat, which can be procured only after about nine years of age; bananas provisioned by humans; and carpenter ants that usually require tool use. Food that needs to be processed includes: *Crematogaster* ant eggs that cannot be eaten until first chewing through brittle, dry branches; hard or thick-skinned fruits like *Strychnos innocua*, *Voacanga lutescens*, and *Saba comorensis*; sugar-cane or arrowroot pith (*Marantochloa leucantha*) that cannot be obtained unless the hard epidermis is chewed through first; and cluster-fruit like *Myrianthus arboreus*, which requires powerful finger strength to pick out its flesh.

All other kinds of food are 'easy food', such as fruits with soft skin and pulp, including figs, or soft, green leaves.

The sharing rate is highly correlated with the solicitation rate (Fig. 3.6). Difficult food is shared more often and it is more likely to be demanded by infants. Given that this is the case, why is 'difficult food' more often shared than 'easy food'? Is it because the infant selects the food and then begs for it, or is it because the mother selects the food that she shares with the infant?

If we examine this paradox closely, we can see that it is because the infant fancies 'difficult food' far more than 'easy food' and not because the mother selects 'difficult food' to share with the infant. So, learning proceeds by an infant's positive interest in new food rather than a mother's teaching tendency, if any.

I define 'food retrieval' (Nishida & Turner 1996) as an infant's recovery of leftovers discarded by its mother. An infant's feeding behaviour begins with food retrieval from its mother. An infant rarely ingests items other than what the mother has eaten or is eating. This enables the infant to learn what to eat from its mother. Food types retrieved are often those that are difficult to process and also are likely to be shared by mothers. Just like the solicitation rate, the retrieval rate reaches a plateau in the second year of life, and gradually declines as an infant grows (Fig. 3.8). However, infants tend to solicit small, difficult food types for sharing, while they often retrieve the remains of large, difficult food types. The function of food sharing and food retrieval lies in an infant's learning food types that it cannot easily obtain or process by itself.

The level of competition for food between mothers and infants remains low throughout infancy (Fig. 3.9). Mothers cease to share the

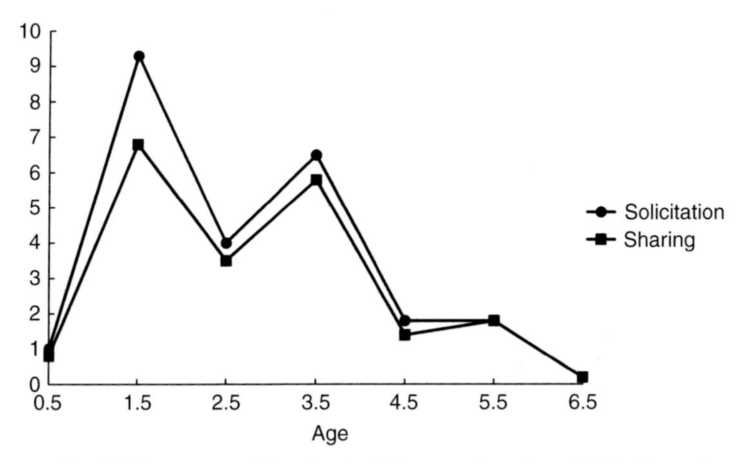

Fig. 3.8 Frequency of begging by infant as a function of infant's age in years. Number of retrieval bouts per number of feeding bouts (from Nishida & Turner 1996).

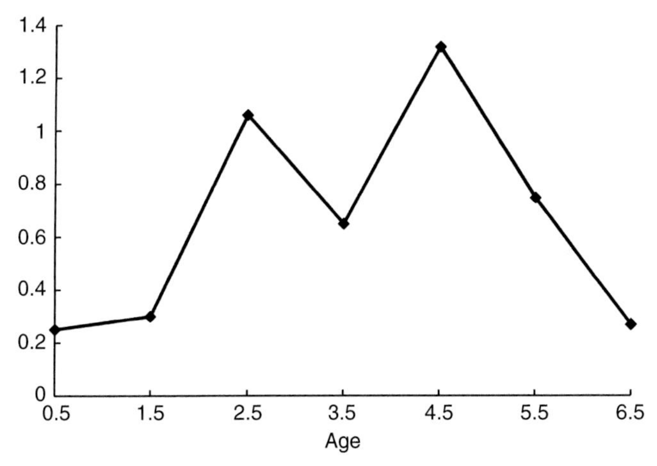

Fig. 3.9 Displacement rate as a function of infant's age in years. Number of displacement bouts per number of feeding bouts (from Nishida & Turner 1996).

same feeding space with their infants as the infants get older. Mothers then become aggressive to them when they approach, so infants gradually increase the distance between themselves and their mothers.

The food repertoire of M-group includes 203 species (340 food items) of plant life, 20 species of vertebrates, and as many as 25 species of insect (Nishida & Uehara 1983; Itoh 2002). Except for meat, a mother seldom shares food with her offspring after its sixth birthday. By the age of three or four years, it is likely that an infant will have basically mastered the group's food repertoire.

3.4.2 Nutritionally trivial, but important in information acquisition

When a mother shares food, she gives only morsels and scraps to her infant, which is no big loss to her. What the infant takes is next to nothing; its nutritional support is negligible. Perhaps one function of mother–offspring food sharing is that by giving the infant information about the assortment of flavours of foods, especially flavours of food that it cannot eat with ease, the mother is unintentionally giving it motivation to eat and to appreciate food. However, there are also exceptions to this rule. An adult female, *Silafu*, often got large quantities of meat redistributed from adult males and in turn routinely shared the wealth with her daughters.

Before we did these detailed analyses, we used to think mother chimpanzees were educating their infants by selecting 'difficult food' and sharing it with them, but actually the infants were actively participating in their own education.

3.5 WEANING

3.5.1 Weaning conflict

Chimpanzee mothers nurse, play with, and carry their infants. They share food, groom and remove lice, ticks, and other debris from their infants, keep them comfortable and warm during rain, and at night share the same bed with them. Other types of care that mothers provide are locomotory aid, protection, and social support.

Locomotory aid is when an infant baulks before a gap in the canopy, and the mother extends a hand or holds fast a branch so the infant can pass safely. Protection is whisking an infant away when it looks as though an aggressive adult male is about to start a charging display or when a dangerous animal comes too near if the infant is playing alone. In social support, the mother takes the infant's side when it is in a quarrel or dispute.

In a broad sense, weaning is the natural extension of the mother's promotion of her infant's independence, from when it is about six months old until it becomes five years old (Nishida 1991a, 1994). A mother tries first to stop carrying the infant and providing locomotory aid, then she stops nursing, and then stops sharing a night bed, which signals that weaning is complete. Other types of care continue until the end of the juvenile period.

Fig. 3.10 The infant prefers to walk in front of, rather than behind, its mother.

Refusal to carry the infant anymore originates with the mother setting out to travel using the 'take off without the passenger' technique. When the infant tries to hop on board, she drops a shoulder or shakes her rump, but if it succeeds in hopping on, she merely sits there encouraging it to walk along on its own. Another tactic to apply when it is time to move on after a rest is for the mother to give the infant's back or rear end a swift push with her hand or foot, making it walk ahead. Infants do not like to walk behind their mothers, but for the most part they are content to walk in front (Fig. 3.10). Perhaps this is because an infant can spot a threat when in front of the mother, while she is the first to notice danger that comes from behind.

In a narrow sense, weaning literally means refusing access to the nipple; it is a clear-cut process. Around the time the infant turns four years old, weaning starts and stops over the next few months. Commencement of weaning may start when the offspring is young, for example, *Sylvie* was weaned at the age of three years.

A mother subtly or overtly assumes positions in which her offspring finds it difficult to gain access to her nipples. Patterns of maternal rejection of her infant vary from individual to individual; she may directly prise the infant from her chest, block access to her chest with her arm, or lie prone on the ground. The first few times the mother denies her nipple to the infant, it just whimpers (vocalising displeasure), but the stricter she becomes, the more its resistance intensifies.

Fig. 3.11 A weanling, *Chopin*, throws a temper tantrum.

A protesting infant will run from the mother, emit screams of upset (Fig. 3.11), roll on the ground, beat its head on the ground, pound itself on the head, and throw a fit. Eventually, the mother almost always gives way to the infant's demands.

3.5.2 Weanling manipulation of the mother

Chimpanzees appear to make great progress in manipulating others from around the time of weaning. Weanlings want milk, attention, and continuing support from the mother. However, the mother resumes oestrus and needs to conceive another child. It is natural that she sometimes forgets her current offspring, who has grown up to juvenility. She now thinks of good mating opportunities, often with specific males, rather than thinking of her offspring. When the weanling notices her lack of attention, it is upset, whimper-screams, and throws temper tantrums. Furthermore, if the mother continues to deny the infant's request for a long period, it sometimes avoids the mother when she finally gives in and approaches the infant to allow nipple contact. The infant's withdrawal of contact may be an expression of anger on its part and appears to resemble the so-called 'withdrawal of reassurance contact', which is an effective tactic

Fig. 3.12 A weanling, *Michio*, rushes to his mother when she shows that she is ready to accept him by extending her arm to him.

in social interactions of adult chimpanzees (see Chapter 8 and Section 8.6.2, in particular, regarding *Ntologi*'s snubbing of *Nsaba*).

When mothers rebuff their requests, infants habitually approach one of the older adolescent or adult males; three infants also approached a human observer, which is potentially dangerous. In front of them, the infant lies supine and cries or screams loudly as in a dangerous position. Two infants even pretended to have been attacked by another chimpanzee, or by a human observer, by circling around and pointing at them with one hand, while screaming noisily (Nishida 1990c). By performing a deliberately risky behaviour, the infant forces its mother to come to its aid. By so doing, a precious opportunity to suckle is gained.

The mother rushes to the scene to rescue the infant, or at least extends her arm (Fig. 3.12). That this is a fiction is shown by the weanling's 'monitoring' the mother's response – glancing at its mother from time to time while it is crying (Fig. 3.13). Thus, its manipulation of the mother's mind is often successful.

The weanling's behaviour is intentional because it occasionally monitors or checks the mother's behaviour, which would otherwise be irrelevant. So far as I have seen, many infants, especially males, show similar behavioural patterns to what the male weanling, *Katabi*, showed during his weaning (Nishida 1990c). This consistency suggests a deeply rooted behaviour that chimpanzees have acquired through

Fig. 3.13 A weanling, *Michio*, monitors his mother's response to his temper tantrum.

natural selection, although the behavioural pattern may be modified through mutual interaction with its mother and other chimpanzees during weaning.

Additional evidence indicates social awareness in weanling chimpanzees. Infants who are denied nipple contact sometimes begin to groom their mothers' ventral surface, gradually coming closer to her nipples, until they finally are allowed to suckle. This tactic was observed in at least four infants at Mahale, as well as at Gombe (Clark 1977). A more subtle difference is the increase in grooming mothers by weanlings. Weanlings rarely groom their mothers when they are not in oestrus. As soon as the mother resumes oestrus, the weanling begins to groom her more often, probably in response to the mother's declining attention to it. Although a mother also increases her grooming in response to earnest grooming by her infant, its rate of grooming still increases during her oestrous period (Fig. 3.14, Fig. 3.15).

This finding supports my assertion that the younger juvenile (or older infant) begins to understand social reciprocity; one must give benefits to a partner in order to gain benefits oneself. Thus, a juvenile female begins to groom an adult female with an infant in order to play with the infant.

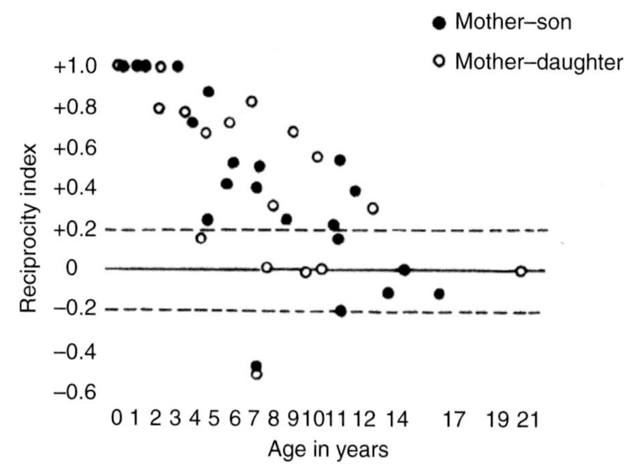

Fig. 3.14 Reciprocity index between mother and offspring as a function of the offspring's age. Reciprocity index by time defined as (A – B) / (A + B), in which A is the duration of the mother grooming her offspring and B is the duration of the offspring grooming the mother (from Nishida 1988).

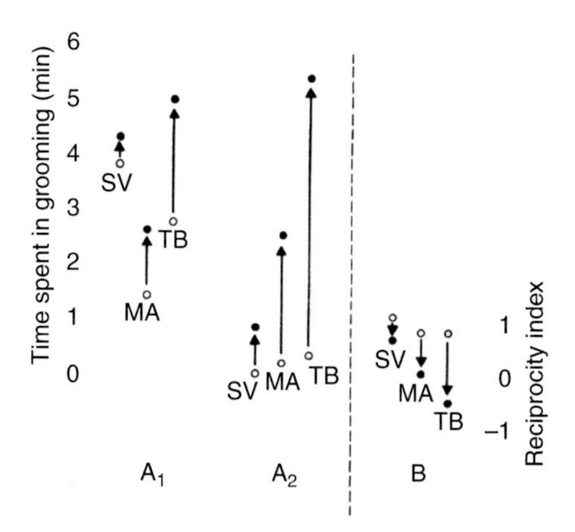

Fig. 3.15 Grooming ratio of mother-to-infant during the weaning period. Maternal grooming per observation hour increased when mothers were in oestrus. A1: mother's grooming time (min) per observation hour; A2: offspring's grooming time (min) per observation hour; B: reciprocity index between mother and offspring. SV is a three-year-old female, and MA and TB are five-year-old males (from Nishida 1992).

3.5.3 Infantile regression syndrome

Breastfeeding during the weaning period is just as irregular as breast-feeding during the newborn period. The stable once-per-hour, two-minutes-for-each-nipple routine collapses, and contact with the nipple and nipple alternation becomes much more frequent and irregular. Both the duration of each suckling bout and the intervals between them vary. For example, maximum bout-length is long in the first year of life, then decreases in later years, then dramatically rises during weaning. This U-shape pattern shows the phenomenon of infantile regression. Extremely prolonged nipple contacts, sometimes lasting as long as 16 minutes, suggest that the mother's breasts are dry and that nipple contact may function only as reassurance. The milk does not flow as well as it used to.

In addition, dorsal riding reverts to ventral riding. This 'infantile regression syndrome' seems to be adaptive 'deception'. It allows the infant to prolong the care its mother gives him, creating the erroneous impression that she began too early to wean the infant. I do not necessarily mean conscious 'deception', but I suggest that an infant who feigns to be like a newborn gains more milk and more protection, thanks to natural selection. On the other hand, 'infantile regression' is really not regression, but is behaviour that only appears as such. As stated above, weanlings understand reciprocal social relations during grooming, so why would they not understand simple bluffing?

An infant apparently throws tantrums intentionally. It purposely puts some distance between itself and the mother and begins bawling with its back to her. However, it does sneak peeks at her from time to time. If the mother does not give it the attention desired, then she just feeds the fire of the infant's anger. It escalates its tantrum. Although the infant is not all that hungry, its fussing will make it appear so. Even if its desire to be held is fulfilled or the soothing nipple is in its mouth but not suckled, the infant may run off at the drop of a hat when friends call it over to play. The infant must have had enough energy to play with age-mates, despite its plaintive solicitation of milk.

An infant's ability to 'trick' the mother goes hand in hand with her ability to see through the infant's deceit and games. If we ask why a mother gives in to her infant's 'mind games', it is because an infant knows better if it is hungry than does a mother (Trivers 1974). If the infant is truly hungry, it goes without saying that for the mother, feeding her infant until it is sated is a smaller cost to pay than starving him; therefore, a mother tends to be lenient with her infant's deceptiveness. And as all wise human parents know, this leniency is not just confined to chimpanzee mother–child relationships.

Around the age of five years, children begin to groom other mothers, in order to get permission to touch or take care of their babies, expressing a desire to be rewarded (Fig. 3.16a, b). During this weaning period, the 'reciprocal mind' seems to bloom into reality (Nishida 1988).

Fig. 3.16a *Penelope* grooms *Chausiku*.

Fig. 3.16b *Penelope* plays with *Chausiku*'s infant, *Chopin*.

In summary, weaning infants apparently revert to neonatal behaviour with frequent whimpering, longer nipple contact, regression to ventral rather than dorsal riding, increased food begging, and so on. However, these regressive behaviours are superficial and should be regarded as an infant's tactics to exploit its mother's care and as a means to pull the mother to its advantage, as Robert Trivers (1974) suggested.

3.5.4 Mating harassment

When a mother's oestrus resumes, a weaning infant continually tries to get in the way of her mating by 'cutting in'; that is, it physically interposes itself (Tutin 1979a). Since the mother is on all fours, her mating partner has to deal with the interfering weanling (Fig. 3.17a, b).

An adult male will do everything in his power to accomplish mating, whether by pushing the infant out of his way with all his might, hitting him, or pinning him to the back of the mother so he is immobile. But the adult male has to be careful, because if he goes too hard at the infant, the mother will stop the mating session. This little episode is intriguing every time you watch it. Sometimes, the session does not come off smoothly, and the adult male grimaces, throws a fit, and sends the infant flying off. The mating session ends abruptly, and the mother and infant threaten the male by barking at him in concert. However, this is an extreme case, and typically copulation succeeds despite a weanling's harassment.

Fig. 3.17a A weanling, *Michio*, throws a temper tantrum.

Fig. 3.17b A weanling, *Michio*, interferes with his mother's mating.

There are sex differences in mating interference, as observed at Gombe (Tutin 1979a). A son tends to protest against his mother's mating more forcefully and obstinately by pushing the male, while a daughter often presents her buttocks to her mother's partner while she is perched on her mother's back.

We might speculate that the weanling is trying to postpone the birth of its younger sibling, so that it may garner more maternal investment (Tutin 1979a). However, it seems unlikely that hindering the mating session will result in any significant delay to a mother's pregnancy, because after all, her infant cannot be on guard every waking moment.

In order to soothe a weanling son's temper tantrum, a mother often solicits mating from him (Fig. 3.18). So far as I could observe during weaning, all weanling sons mated with their mothers during her oestrus. This son–mother mating seems not to occur elsewhere,[1] so this reassurance mating may be a custom at Mahale – a social custom almost beyond our imagination!

About the time the child turns five years old, the intense mother–offspring conflict ends. The weanling has mastered the group's food repertoire, and even if there is a deluge, he has grown so much that

[1] That this does not occur at Gombe is obvious, because there is no description in Tutin (1979a) or Goodall (1986). At Ngogo, John Mitani has never seen it (pers. comm. May 2010).

Fig. 3.18 A weanling son, *Michio*, mates with his mother.

it is unlikely he would suffer by not being snug next to his mother's bosom. In this way, before long the mother–offspring relationship resumes its once peaceful status.

3.6 JUVENILITY

3.6.1 Variations of play and harassment

During the juvenile period, the mother no longer carries her offspring; rather, the offspring follows its mother. A downpour does not mean it needs to be at its mother's bosom, as its body is now big enough to thermoregulate. At night the juvenile sleeps in a bed of its own making.

At Mahale, the age of five years marks the beginning of the juvenile stage and coincides with the time when a chimpanzee can process and eat the important *Saba comorensis* fruit unaided. *Saba comorensis* is the single most important food of M-group chimpanzees, being annually available from August to January. Juveniles must master how to process this fruit to ensure their survival. However, at age five, many food items such as the thick-skinned *Voacanga* still present a challenge and cannot be eaten skilfully. Juveniles are not able to handle these kinds of food until they are six or seven years old. Though still a bit clumsy, they muster up the patience to become adequate fishers of carpenter ants, although they began to fish for ants at the age of three or four. The advanced juveniles make their first successful hunt at the age of eight. A male child, *Linta*, made his first catch of a blue duiker's fawn close to his eighth birthday, but an adult male soon snatched away the prey. The ability to catch colobus monkeys comes later, usually during early adolescence.

For both males and females, the juvenile stage is by far the most active part of a chimpanzee's life, a time overflowing with curiosity and wonder. They love playing, and when marching before or after mother they purposely turn somersaults and pirouettes, or show a leaf-pile pulling performance (see Chapter 4). They stray from the path, kick or slap a few times at a buttress root, or pick up and throw a hard-skin fruit or small stone. Down by the riverside, juvenile males mimic the adult males by throwing stones; juvenile females play gymnastics in the sand.

There is also a wide variety of social play. During infancy, a kind of jumping-jack game in the treetops is popular, but during the juvenile stage, wrestling on the ground and playing tag are the main events. Twirling (or circling) round and round a tree or a seated mother (or sometimes with no axis) is popular not only among juveniles, but also between juveniles and adults and even among adults only. Wrestling more often occupies juvenile males than females (Hayaki 1985).

Splashing about in the water is also a big event for juveniles. Activities include turning over stones underwater, splashing with a friend, or slapping the water surface with a branch. Oddly enough, during the latter part of adolescence, individuals discontinue playing in the water and stay as far away from it as possible.

Some behaviour is a combination of play and assault and can be referred to as bullying or 'harassment'. Juveniles, especially males, approach another youngster or adult and begin wielding a branch, trying to startle them, throw sticks at them, or slap them and then run away. Harassment that emerges in the early years seems to be behaviour in which one individual tests the boundaries and reactions of another individual (Adang 1984).

Once, a seven-year old male, *Carter*, approached the alpha male, *Ntologi*, and with one hand began shaking a fallen tree up and down, startling *Ntologi*. *Ntologi* was furious and aggressively chased the fleeing *Carter*. It seems that youngsters observe their counterparts' reactions to these provocations and from this amass information on a group member's temperament. *Carter* never provoked *Ntologi* again.

The most common pattern of harassment is a juvenile or adolescent male harassing an adult or adolescent female (Nishida 2003a). An individual who is harassed displays all sorts of counter-behaviour:

ignore the perpetrator; fly into a counter-attacking rage by barking at or chasing him; or scream and run away. The ultimate purpose of the male youngster's harassment is to dominate targeted females by making them pant-grunt to him. An adult male's dominance of all adult females makes it easy for him to mate with all the females without female resistance, if a more dominant male is absent.

3.6.2 Juvenile behaviour and sex differences

Adult males, as well as young play partners, fascinate juvenile males. But a mother with a juvenile's baby brother or baby sister in tow has no attraction to an adult male. After a brief greeting and grooming with the adult males, it is more advantageous for a mother to part ways and forage with her family. This situation is when a mother and her juvenile's views split.

But in the end, between his sixth and seventh year, the juvenile has no alternative but to follow and to obey his mother. This is the period when juveniles often lose their way; adult males or friends distract them, and when his mother is departing, a young male may lose track of her. When he notices that the mother is nowhere to be found, he becomes agitated, begins to whimper, and if he still cannot find her, then he scales a tree and screams repeatedly. I have seen a juvenile continue to scream non-stop for two hours in the treetops.

However, when males are more than eight or nine years of age, separating from the mother and tagging along with a group of adult males becomes more regular. If a juvenile male has an adolescent or adult older brother, he tends to become independent from his mother sooner.

Compared with a juvenile male, a juvenile female clearly has a much stronger tendency to follow after her mother. This is partly because a juvenile female is more attracted to her younger siblings than is a juvenile male. Also, a juvenile male is more attracted to adult males and playmates than is his female counterpart. Still, for female youngsters the opportunity to roam with large groups that include adult males also means an abundance of playmates. In these contexts, if mother and daughter disagree about the direction in which to roam, when her mother is not looking, the daughter carries away her younger sibling on her belly and follows after the big party including adult males. Now the mother is a 'hostage' who reluctantly treads on her daughter's heels. This social tactic allows a daughter to manipulate her mother.

As a mother advances in years, she gives birth less often. The last offspring to be weaned is delayed in both weaning and psychological independence. There were two old mothers, *Happy* and *Wakasunga*, and their youngest children, *Hanby* and *Linta*, respectively. We became very concerned as to when these two sons would finally become independent of their mothers. As it turned out, *Hanby* at the age of 12 years still clung to his mother when she died, and *Linta* was still following his mother when he caught cold and died at the age of eight years. Pampering the 'last baby' in a family is something that both chimpanzee parents and human parents share. Having no necessity to invest in the next offspring, it stands to reason that a mother does everything in her power to give care and affection to her final offspring.

3.7 ADOLESCENCE

3.7.1 The adolescent male

Male chimpanzees experience a period when their scrotum rapidly enlarges and becomes highly visible. This rapid physical transformation usually occurs around nine years of age and shocks us human observers. At Mahale, I have been able to confirm the age of first ejaculation for only three males. Their respective ages were nine, nine and three months, and ten years old (Nishida *et al.* 1990). Thus, the male chimpanzee's adolescence is from nine years of age when he can ejaculate until about age 15 or 16, when his body has reached adult size.

A visual feature separating juvenility from adolescence is the loss of rump hair. The only exceptions were two males, *Bembe*, who even at age 18 years had a little rump hair remaining, and *Bonobo*, who maintained it even at age 20.

Upon entering adolescence, both males and females must depend on themselves for subsistence and protection. They can handle and process any kind of food, and from the time they are about 10–11 years old, they acquire the ability to catch colobus monkeys.

Males who have entered adolescence do not accompany their mothers any more; instead, they travel with adult males. When encountering an adult male, the adolescent male lets out a series of 'pant-grunts', as if he has gone completely mad. In other words, he stands on all fours and nods his head up and down while giving out intense panting sounds like 'Aha, aha, aha'. This is the tone of greeting voice that an individual of lower rank uses to an individual of higher rank. An adolescent male will ask like this for an adult male's tolerance or permission, but it usually is not successful. Adult males treat juvenile males with generosity, but once the juvenile becomes an adolescent, the positive treatment they once received alters dramatically. Sexually mature adolescent males have grown up to be sexual rivals for adult males. When an adolescent male comes near for greeting or salutation, the adult male does quite the opposite, sometimes chases him, slaps or stamps on him, or gives him a nasty bite.

In this way, the adolescent male comes to occupy a peripheral position in the group (Goodall 1986; Pusey 1990). However, adolescent male chimpanzees do not form all-male groups, as seen among macaques, langurs, or gorillas. Why adolescent male chimpanzees do not form all-male groups while many other non-human primates do is an intriguing question that awaits explanation. Perhaps interest in adult males overwhelms their interest in peers? Social status is already determined between adolescent males, but instances of emitting pant-grunts to one another are rarely, if ever, witnessed, except towards the older adolescent males who are approaching adulthood (Hayaki *et al.* 1989). To refrain whenever possible from pant-grunting appears to be an essential male strategy throughout adolescence and adulthood.

In moving from adolescence to adulthood, a male chimpanzee regularly shows a kind of display behaviour called harassment (see above). He provokes and attempts to intimidate adult females. This behaviour develops from the harassment seen in juvenile males. He thrashes a thick vine in their faces or throws a stick, threatening them. This goes on until the female submissively greets him with a pant-grunt. Screaming is not enough, because the female may expect some adult male to come to her rescue if she screams loudly. Screaming is never a genuine sign of submission but rather one of resistance. Using this behaviour, an adolescent male becomes dominant of all females, although his mother may not be afraid of her grown son. Even when sons become adult, most mothers do not pant-grunt to them.

Another behaviour peculiar to adolescent males is to select a middle-aged adult male of comparatively low rank and accompany him. In chimpanzee society, there is something like an 'all-males club' for adults, to which it is difficult to acquire admission, like an English men's club that young gentlemen find difficult to join. If you do not have membership privileges, you are not allowed to participate in a grooming party or a meat-eating cluster composed mainly of adult males and adult females. An early step along the way to becoming a club member is to befriend one of the less aggressive, low-ranking adult males.

Alternatively, an adolescent male may identify the lowest-ranking adult male and relentlessly terrorise and harass him until he admits defeat, whereby the two reverse roles and rank. Now the adolescent has become a 'young adult male', but several more years are needed to become a true member of the adult male club.

The age at which a male can make all females acknowledge his dominance depends on the male in question. Hitoshige Hayaki surmised that an adolescent male who defeats the lowest-ranking adult male, and becomes stronger than any adult females at an early age, will later achieve high-ranking status. I share this opinion with him. For example, *Nsaba* dominated all the females at age 12, while small-sized *Darwin* found it difficult to make all females pant-grunt to him even at age 17.

3.7.2 The adolescent female

An adolescent female shows physiological transformations that are the result of oestrus. From about the age of eight years, her sexual skin begins to swell up, and by age ten years it swells maximally and mating commences. It is reasonable to say that the female's adolescence begins at age nine years. However, it is more difficult to put a finger on exactly when her adolescence concludes. At Mahale, the majority of individuals experience their first births at age 15 (Hiraiwa-Hasegawa *et al.* 1984; Nishida *et al.* 1990, 2003), so pregnancy begins at age 14. It would be reasonable to say that adolescence continues until 13 years of age.

Around the time of the first oestrus, females copulate incessantly, but most of the copulation is with juveniles and is almost playful. Females love to muck around too, so having a juvenile male as a mating partner almost always turns into a wrestling match. Adolescent females also copulate with adult males, but the phenomenon of 'adolescent infertility' (Tutin 1979b; Hasegawa & Hiraiwa-Hasegawa 1983, 1990) is at play here, such that even if the adolescent female copulates, she does not become pregnant. When an adolescent female encounters an adult male she gets excited and salutes him by pant-grunting repeatedly. Although she is an adolescent, the adult male does not treat her as harshly as he does an adolescent male.

The behaviour that stands out most clearly with adolescent females is 'alloparental' or babysitting behaviour (Nishida 1983b). If the mother of, for example, a two-year-old infant is relatively tolerant, she allows the adolescent female to hold her infant and care for it.

Fig. 3.19 A mother, *Wakampompo*, allows her daughter, *Wantumpa*, to touch her young brother, *Bonobo*.

The mother is more tolerant of her own adolescent daughter and will allow her to handle a baby of under one year of age (Fig. 3.19). If the mother permits, the adolescent female has the baby all to herself, away from the mother's side. On such occasions, the adolescent female sometimes gets herself into a fix, because the baby begins to squirm and fuss, wanting to nurse, but the mother is nowhere in sight. As the baby's whimper begins to intensify, the adolescent female loses confidence in keeping her charge, running hither and thither hunting for the mother. If the adolescent caretaker has trouble tracking down the mother, the baby tries to flee from her. The babysitter then grimaces, lets out a whimper, and even throws a tantrum herself!

When an adolescent female's oestrus begins, she distances herself from her mother and starts accompanying adult males. She returns to her mother after oestrus has ceased. As at Gombe, adolescent females in Mahale have a much longer-lasting and intimate relationship with their mothers than do adolescent males (Pusey 1990; Nishida 1994). And yet, at the age of 10–12 years, an adolescent female immigrates into other group (community) ranges (Fig. 3.20). It is not as though anyone drives her away; she emigrates entirely of her own accord.

The adolescent female shows many kinds of social curiosity towards immature individuals. She may throw a stone at an adolescent female peer in order to watch her response. She may find 'teasing' a

Fig. 3.20 The K-group-born female, *Nkombo*, immigrated into M-group at 11 years old.

juvenile fun. For example, the ten-year-old female *Tula* 'fished' the four-year-old male, *Linta*. One day, *Tula* picked up the long pelt of a colobus monkey discarded by an adult chimpanzee. Perhaps it was a tail or strip of back fur. *Linta* really wanted that pelt, as many youngsters do. When he began to nag *Tula* for it, the latter climbed up a tree and dangled the pelt, coaxing *Linta*. Every time *Linta*, who was on the ground below, jumped to try to grab the pelt, *Tula* would jerk it up and then hang it down, coaxing again and again, making a fool of *Linta*. But at last, *Linta* leapt up and when he grabbed onto a corner of the pelt, *Tula* let out a dreadful shriek, leaving me in fits of laughter (Nishida 1994).

I believe this is evidence of a wild chimpanzee reading another's mind and making fun of it. Although this is my hypothesis, I cannot hit upon any other parsimonious hypothesis to explain this episode. Like *Nsaba*'s case, this episode is characterised by 'rare' behaviour. *Tula* dangled the pelt from above. This is unusual. Every time *Linta* jumped, Tula jerked it up. This is also unusual. The fact that *Tula* screamed when *Linta* grabbed the fur suggests that she did not want to give the pelt to *Linta*. This sequence of events indicates that chimpanzees engage in something like human teasing or ridicule. It is difficult to study the internal states of animals, especially in the field. However, careful observation sometimes provides the window through which we can reasonably infer their cognitive abilities.

3.8 SIBLINGS

3.8.1 Alloparental care for young brothers and sisters

As noted earlier, female chimpanzees most often give birth for the first time at age 15. It takes about five years to wean a baby chimpanzee; the gestation period is eight months, and after weaning the previous offspring and resuming oestrus, it takes four months to conceive, plus four years that are dedicated to the next child's infancy. So, if a female has her first birth at age 15, by the time that child has been weaned, the mother will be 20 years old and will have the next offspring at approximately 21 years of age. Following this pattern of six-year intervals, she will have her third child at 27 and her fourth at 33. The final delivery may be at the age of 39 or 45. If she does not lose any baby, she will have a maximum of six offspring.

Thus, unlike many contemporary human societies, chimpanzee siblings have, on average, six years between them. Additionally, the likelihood of having half-brothers and half-sisters, rather than full sibs, is high, as females are promiscuous, and alpha males, who have a high probability to be their sex partners, change every 5–6 years on average (Chapter 8). This suggests that chimpanzee sibs are expected to be less cooperative than human sibs. Recent DNA paternity analysis conducted at Ngogo indicates that this is the case (Langergraber *et al.* 2007).

A commonly seen grouping is when the elder offspring is eight or nine years old, and the younger offspring is one or two years old, and the older sibling almost always takes on the role of babysitter. Transport, grooming, play, protection, and so on are a typical day's duties for big brother or big sister. When the baby is in her earliest days, the mother is reluctant to pass the responsibility off to the older sibling. When *Sylvie*'s sister was born, she could not wait to touch and hold her, but their mother, *Silafu*, would not allow it. *Sylvie* then devised a plan to get her wish; as she was grooming her mother's back she extended a stealthy leg to touch the baby. This tactic was later confirmed for other older sister–baby sibling pairs, who hit upon the same idea!

Most mothers permit the elder offspring to carry the baby after the infant is more than one year old. The mother travels at the front, while the big brother or sister carries the baby on their belly. But when startled, the big brother or sister jumps onto the mother's back, baby and all! If an elder sibling has an accelerated weaning, the mother carries the former on her back and the new baby on her belly (Fig. 3.21).

Fig. 3.21 A mother, *Wakampompo*, simultaneously transports both older and younger offspring.

3.8.2 Various sibling relationships

There are all sorts of sibling relationships. *Abi* had two big brothers who were born seven years apart and got along well. They were almost inseparable, often wrestling and mucking about. After their mother died, seven-year-old *Abi* spent more of her time with her second-oldest brother, *Toshibo*. Two years later, after the eldest brother went missing, *Abi* accompanied *Toshibo*, often being groomed by him. But usually *Abi* was obliged to groom *Toshibo* first (Fig. 3.22).

When her mother died, *Shinako* was seven years old. She was always tagging along with her big brother, *Ryo*, who was seven years her senior. Within the group, the individual she groomed the most was *Ryo*; *Shinako* and *Ryo* had a similar arrangement to *Abi* and *Toshibo*, which was 'I first groom you, so you groom me'. However, *Shinako* groomed *Ryo* far more than he groomed her.

Yet, there are also cold and distant sibling relationships. When her mother *Wantendele* died, even though *Maggie* was only three years old, her big brother, *Masudi*, ten years her senior, carried her on his back only for the first day. Instead, *Maggie* was adopted by an adolescent female, *Tula* (see Chapter 6). Such a cool sibling relationship is not always due to age-gaps between siblings. The case of a big brother becoming independent from his mother at an early age produces the

Fig. 3.22 *Abi* grooms her brother, *Toshibo*.

same result as a large age gap. *Nsaba* stopped spending time with his mother at the early age of eight. He did not associate with his little sister, *Tula*, who was seven years his junior. Even having seen them many times, you would have no reason to believe they were siblings.

In M-group, the largest family by far was the *Wansonbo* family. There were four children, *Kat* (elder daughter), *Bembe* (elder son), *Bunde* (second daughter), and *Bongo* (second son). The reason behind their unusual size was that, curiously, *Wansonbo*'s daughters never transferred to another group. Since *Kat* continued to accompany her mother, even after giving birth, their little family swelled to seven members. This massive family mostly roamed in the south sector of M-group's range, always doing everything together, whether it was grooming, playing, or babysitting.

Meeting up with them was a pleasure for me. *Wansonbo* was a loving mother but a bit of a blockhead. Whenever she fished for carpenter ants, she used a probe that was far too flimsy and never worked. Then, she took her six-year-old son *Bongo*'s probe (he was sitting beside her), tried it out, and it was a huge success! It reminded me of the Japanese saying, 'Taught by a child whom one carries on one's back.' This observation suggests that an infant does not always learn the fishing technique from its mother.

This family had its share of misfortune, and *Wansonbo*'s later years were lonely. First, her eight-year-old son *Bongo* was taken by a lion. Then, her 'tardy to tie the knot' 14-year-old daughter *Bunde* became ill. Her coat lost its lustre, her health declined rapidly, and she died. After *Bongo* there was a little sister, *Binti*, but she died when she was just two years old.

Kat had a second daughter in 1988, when her eldest *Totzy* was six years old. In the early 1990s, *Kat* and her offspring often joined in a big party including adult males, after being lured by the spirited *Totzy*. *Kat*'s family and the elderly *Wansonbo* are not seen together as much as they used to be.

The 20-year-old *Bembe* is a tough guy on his own turf but is hardly recognisable when on his own. He pant-hoots loudly and stages intense displays. When he does, his mother and *Kat* and her offspring emerge from somewhere to watch, and then they all go off and forage in a fruit tree that *Bembe* has found. *Kat* and *Bembe* are a rare example of a sibling relationship that continues to be intimate even into adulthood.

3.8.3 The *Opal* family

Opal immigrated into M-group in 1983 and occupied a core area at the northern part of M-group's territory. She gave birth to her first daughter, *Ruby*, in 1985, first son, *Orion*, in 1990, and second son, *Oscar*, in 1996. Like *Kat* and *Totzy*, *Ruby* did not emigrate from M-group but instead stayed to give birth to her first daughter, *Rubicon*, at the young age of ten years. Thus, *Opal* began to enjoy the largest family since *Wansonbo*'s family had disappeared.

In 1999, the infant *Rubicon* was attacked by the two highest-ranking males, *Fanana* and *Kalunde*. *Opal* helped *Ruby* rescue *Rubicon*, and the mother–daughter coalition succeeded in helping *Rubicon* to recover, although *Rubicon* lost some fingers and toes that were badly bitten by the males. This was an unsuccessful infanticide (Sakamaki *et al.* 2001). After the incident, *Opal*, *Ruby*, and their offspring made a more tight-knit team and moved together most of the time. *Opal* became very sensitive to her granddaughter, perhaps because of the damage to the baby's digits. As *Rubicon* grew up, she engaged in social play in the trees, chasing and hanging with one hand and pushing her playmate down to the ground with the other. On such occasions, it was *Rubicon* who was forced to fall down most of the time. In spite of this misfortune, *Rubicon* never showed dissatisfaction in any facial expression or gesture. *Opal* was often accompanied by *Oscar* and *Rubicon*, and strangely *Ruby* was nonchalant even when her small daughter was out of sight. One day, when *Rubicon* fell into a hole in the ground and screamed, *Opal* rushed to the scene earlier than *Ruby*, although both of them were the same distance from *Rubicon*. Thus, the grandmother became *Rubicon*'s favourite, rather than her mother.

Sibling relationships are variable, from very intimate to cold. However, mother–daughter relationships are always warm, and they may form strong coalitions. This confers a great advantage on a female who does not emigrate. In spite of this, more than 90 per cent of females emigrate from their natal group and immigrate into a group where no maternal support is available. I will discuss why in the next chapter.

3.9 AGEING AND DEATH

3.9.1 Age of old females

It is accepted that a chimpanzee's life span may approach or exceed the age of 50 years. But how do we know the age of females? I will use *Fatuma* and *Wakusi* as examples. I first saw *Fatuma* as an adolescent female in 1973, when she immigrated into K-group. So, it is likely that her age in 1973 was 11. As 37 years have passed since then, she was 48 years old in 2010. When *Wakusi* was identified in 1980, she was always followed by a juvenile female, *Kantamba*, who was probably three years old. If *Kantamba* was her first offspring, and since the age for first births at Mahale is around 15, we can work out her age as 48 (3 years + 15 years + 30 years). Of course, this may be too conservative, because she may have had another offspring before *Kantamba*. Although neither *Wakusi* nor *Fatuma* did, some old females, such as *Wanaguma* and *Wakasunga*, begin to suffer from cataracts with white and cloudy eyes in their early forties.

As of 2010, M-group had four females over 45 years old. We were surprised to see that females over 40 years, such as *Wakusi*, *Calliope*, and *Fatuma*, still gave birth. (At Gombe, *Fifi* gave birth to her seventh offspring at age 45.) This depends on a mother being blessed with good health; a mother in declining health trying to wean the fifth child is in a precarious situation with high risks. It is usually impossible for a mother to wean an offspring that she bore at around the age of 45. Therefore, it is fortunate if a mother can successfully rear five children in a lifetime. Since adolescent females emigrate from their natal range at around 11 years, if the eldest child is a female and there are four children, they will not live together in the same group.

Our long-term research shows that female chimpanzees have relatively few post-reproductive years and that menopause, accordingly, is a unique human characteristic not shared by other great apes (Nishida *et al.* 2003; Emery-Thompson *et al.* 2007).

3.9.2 The oldest female

Waganuma, who died in 1990, was the spitting image of the alpha male, *Ntologi*. What is more, they were together most of the time, and because she was often groomed by *Ntologi*, we inferred that she was his mother. Kenji Kawanaka first identified and named him in 1972. The first time I laid eyes on *Ntologi* was in 1973, when he was a young adult male. I presumed *Ntologi* was *Waganuma*'s firstborn at 15 years of age, and taking her age at the time and adding another 15 years as an estimate of *Ntologi*'s age the first time I observed him, plus another 17 years up to the year *Waganuma* died, her life span was 47 years. This was a conservative estimate of her age.

In her later years, she had cataracts, her weight dropped, her hair thinned, she had a hunched back, and she found it hard to keep up with her group mates when on the move. The year before she died, it became impossible for her to roam with her M-group associates, including her son. She continued to live a life in the south sector of M-group's range, where she enjoyed her associates' company only when they decided to visit. Even at tree climbing she was handicapped, and when she swung from branch to branch, I caught myself subconsciously calling out, 'atta girl!' to her (Fig. 3.23).

Fig. 3.23 *Waganuma*, *Ntologi*'s mother, in a tree.

She may be the oldest chimpanzee with whom we have been acquainted. I never saw her give birth, and for at least the last four years of her life, she displayed no swollen sexual skin. This is not definitive proof that chimpanzees experience menopause, because four years is too short in comparison with 30–50 years of human female menopause.

3.9.3 Old males

We have no obvious signpost for estimating the age of a male who is advanced in years. I know of about 15 males who are, or were, obviously more than 36 years of age. With these males, too, their nomadic range is often contracted and their ability to keep company with associates has dropped. Nevertheless, most elderly males enjoy popularity with the high-status, middle-aged, or young adult males, and when they encounter each other the elderly males receive grooming. When there is an opportunity to eat meat, elderly males have privileged access to prime pieces. Aside from diminished physical strength, most elderly males are equipped with what one might call influential power.

Elderly females are often accompanied by adolescents and juveniles. They are especially adored by and sought out by orphans. We find that when humans get along in years, they are not totally unlike chimpanzees; the men become engrossed in politics and the women thrive on caring for their grandchildren.

3.9.4 Dealing with death

Frequently I am asked questions about chimpanzee death like: ‘Do chimpanzees bury the corpses of their group members ?’ Or, ‘Are there graves?’ These sorts of behaviours have never been seen. Indeed, only rarely do we have the opportunity of seeing a chimpanzee’s corpse.

On these rare occasions, freshly dead chimpanzees are always discovered on well-travelled animal trails or in the heart of ravines. Chimpanzees who are ill and whose bodies gradually become emaciated, or those who die of natural causes (e.g. old age), lag behind their associates. When they eventually die, it is usually away from the trail. At Mahale, the bushpigs usually devour corpses. After this occurs, the only remains you may find are the skull, lower jaw, pelvis, and femur.

Mothers carry their deceased offspring, so we often catch sight of their bodies. The infant mortality rate is very high, with 60 per cent of chimpanzee deaths occurring before the weaning stage (Fig. 3.24).

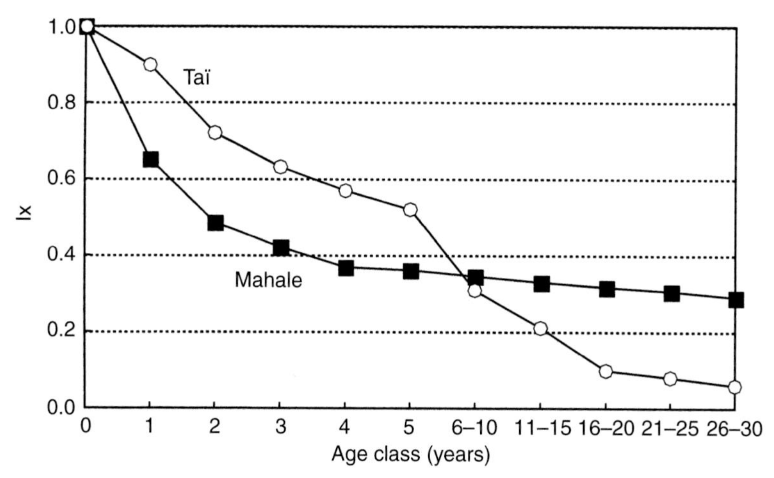

Fig. 3.24 Survivorship curves for Mahale and Taï chimpanzees (from Nishida *et al.* 2003). Taï information from Boesch & Boesch-Achermann (2000)

There are more than 60 parous females on the registry for M-group, but only four are confirmed as not having suffered the grievous experience of losing a child; only one of these was an elderly female. She was the alpha female, *Wakampompo*, who gave birth to her eldest daughter in 1976, her eldest son (*Bonobo*) in 1981, her second daughter in 1987, and her second son in 1992.

Wakasunga gave birth to eight children, but she weaned only two sons, *Lulemiyo* and *Linta*. However, *Lulemiyo* disappeared at age ten, and the youngest, *Linta*, died at age nine. At age 40, *Wakasunga* died a lonely death from emaciation.

Mothers carry their deceased offspring clutched in one hand or on their backs. When the corpse begins to rot, they place it on the ground and appear overwhelmed as they swat flies away from the lifeless baby. Dead infants are carried for a few days to one week, but I once saw an elderly female carry her two-year-old dead son for over three months. The younger the mother, or the younger the baby is, the more there is a tendency to dispose quickly of the corpse. They usually desert a dead newborn within a week but continue to carry 1–3-year-old infants for several weeks or longer. This contrasts sharply with Japanese macaque mothers, who tend to desert older infants earlier than younger ones (Sugiyama 2009). This suggests that the decision process to abandon the corpse is different. A macaque mother's decision may depend on the weight of the dead infant, while a chimpanzee mother's decision may depend on the degree of affection for the dead infant.

There is another important factor influencing the length of time a dead infant is carried. If the time of death is during the rainy season, the corpse is quickly disposed of. This is because the corpse rots quickly, becomes dismembered, and is impossible to carry. During the dry season, however, as the corpse mummifies, it can be carried more easily.

The first encounter with a corpse is met by curiosity from other chimpanzees, who stare and move close, trying to touch it and smell it. Juveniles display a much stronger curiosity about the corpse than do adults. But after a few days have passed, everyone goes about his or her business, no longer taking specific interest in the corpse.

4

Play and exploration

Play versus exploration: how can we distinguish them from each other? This is difficult because their form and contexts appear similar. Exploration comprises behavioural patterns that animals seem to use for collecting information on the surrounding environment, an endeavour fuelled by curiosity. Exploration is often intermingled with play, in particular in immature chimpanzees, as described below. Accordingly, I do not focus on differentiating them or subject the reader to a dry disentanglement of the two 'categories' in detail. Instead, I simply emphasise that both exploration and play seem to be seeds of culture. Through video monitoring of Mahale chimpanzee behaviour (Nishida *et al.* 2010), we have found that there are many innovative patterns in play.

4.1 EXPLORATION

4.1.1 Curiosity and the collection of information

Chimpanzees inspect any unusual object in their environment. If we leave binoculars, a handkerchief, field notes, or anything new to the environment behind on the observation path, one of the chimpanzees, usually a youngster, picks up, sniffs, and sometimes even carries the foreign object.

One day, when I left a *panga* (a long bush knife or machete) on the trail while writing, a juvenile female, *Abi*, picked it up then threw it down. She was surprised to hear the sound of it hitting the ground. I imagined that if an adult male threw this, it could be extremely dangerous! An adolescent male, *Darwin*, carried a cap my field assistant took off while resting. Abandoned village sites are a treasure trove of toys such as earthenware and pots, which juveniles and infants often pick up and manipulate.

Older infants and juveniles tend to enter any conspicuous hole in the ground, such as those dug by an aardvark, to see what is inside. They also investigate the crevices of rocks, entrances of ant nests, and every nook and cranny in trees. During such exploration, they typically enjoy doing somersaults, sliding, slapping, stamping, and various other social play patterns. Among youngsters, therefore, exploration and play are often intertwined and thus difficult for us to discern.

4.1.2 Chimpanzees abuse, kill, and play with a leopard cub

In 1984 Mariko Hiraiwa-Hasegawa and colleagues watched chimpanzees entering a rocky cavity to search for a leopard cub. Finally, an adult male roused the cub, then smashed it to the ground, killing it. No mother leopard was seen. A juvenile male and an adolescent female chimpanzee treated it as a stuffed doll. The juvenile male climbed a tree with it and dropped it repeatedly. The adolescent female made a bed for it in a tree and showed 'alloparenting', including grooming the dead cub for a long time (Hiraiwa-Hasegawa *et al.* 1986; Byrne & Byrne 1988). Here, also, we find exploration and play intimately interrelated.

4.1.3 Responses to fresh carcass of adult female leopard

On 8 August 1999, when I followed a young adult male, *Hanby*, we came across the freshly dead body of a female leopard (Nishida & Zamma 2000; Nishida, unpublished data). She had a fatal wound in her neck and appeared to have been killed by another leopard. *Hanby* hit the carcass then rushed to call his associates. In response to *Hanby*'s fearful call, first two adult males, *Alofu* and *Masudi*, quickly arrived and repeatedly slapped the dead body with one or both hands. Then, all of the chimpanzees who were scattered around the area began to gather there. Chimpanzees touched, sniffed, slapped, and kicked the carcass. They also pulled, dragged, pushed, tried to lift, and dropped this dead leopard. They investigated the earlobes, picked at hair, inspected paws, whiskers, and head, nibbled the skin, and even groomed the body

Fig. 4.1 An adult male slaps at a fresh leopard carcass.

(Fig. 4.1). Most of these actions seemed to be aimed at collecting information: touch and slap to ascertain whether the body was dead or alive; drag, pull, and lift the carcass to weigh it; inspect the paw to study the predator's weapons; and nibble to evaluate the edibility. Some chimpanzees (mostly youngsters) did not approach the carcass on the ground, but instead climbed a tree overlooking the leopard. They threw many slender branches at the carcass, as well as shaking branches at it. However, no one struck the leopard with a stout stick as was induced by Adriaan Kortlandt for the chimpanzees of West Africa (Kortlandt & Kooij 1963; Kortlandt 1967). The entire display continued for more than six hours; some remained there the whole time without eating, but many left to look for snacks before returning to the leopard. Finally, a few females began to nibble the flesh, but never to the extent of relishing the meat, and one adolescent female, *Pipi*, even pulled the entrails from the belly and hung them in the tree canopy like an electric wire (Nishida & Zamma 2000; Nishida unpublished data).

This episode shows us that chimpanzees will grasp any opportunity to obtain important information, in this case on a major predator. They seem to think that it would be a crime to pass up a chance to explore such a rare find. After making such a find, they even postpone foraging and eating. A chimpanzee has as much curiosity as a human. The members of M-group took advantage of this one rare opportunity, and seized the chance to obtain hands-on knowledge of leopards.

Curiosity trumped appetite, because information on leopards was likely to be more important than a single day's full stomach. For long-lived animals such as chimpanzees, it is sometimes more important to gather information, which takes priority even over daily subsistence activities such as foraging. The leopard is a natural enemy, and the information obtained on this species may well be worth going on a six-hour fast.

4.1.4 Leaf-grooming as a lice-removing technique

During or after social grooming or self-grooming, chimpanzees at Mahale and Gombe take a leaf, bring it to the lips, and fold it with the fingers. Then, they squash the folded leaf with their fingers and drop it. They sometimes attend fixedly to this business and repeat the sequence more than five times. Often, one or a few chimpanzees approach and peer at this leaf-grooming. Leaf-grooming was first described by Jane Goodall as early as 1968 (Goodall 1968), but no one has really understood the function of this behaviour until recently.

In 1999 Koichiro Zamma was studying grooming behaviour in terms of its hygienic function. One day when he was observing an adult male, *Masudi*, leaf-grooming, he noticed a black spot on the leaf that *Masudi* was manipulating. Usually, the chimpanzee eats up such a 'black spot' and nothing remains. Fortunately, Zamma recovered the black spot on the leaf that *Masudi* discarded: it was a louse (*Pediculus schaefi*). It turns out that when chimpanzees put their lips to a leaf, they are placing a louse on it, and by folding the leaf with a hand they completely crush the louse. Thus, the mystery of a chimpanzee behavioural pattern was solved after more than 40 years of research (Zamma 2002)!

Leaf-grooming develops slowly. The two youngest chimpanzees observed to manipulate a leaf were two years old, but it was not in the usual context of social grooming. From 1999 to 2004, I observed 11 chimpanzees for 1.5 months per year, during their entire period of infancy and early juvenility, and I saw fully fledged leaf-grooming for the first time between the ages of three and six years old, with the average being 4.5 years (Nishida, unpublished data).

At Bossou, chimpanzees place a parasite on the palm of one hand and then smash it with the index finger of the other hand. Similarly, chimpanzees of the Taï forest smash parasites against the forearm, using the index finger of the opposite hand (Nakamura 2010). Thus, similar behaviour with a similar function can vary between study sites.

4.1.5 Why do so many gather?

A chimpanzee often looks into another individual's face from a distance of a few centimetres, while standing quadrupedally, when the other is eating, leaf-grooming, or investigating or licking a wound. We call this 'peering'. Peering seems to be one of the remarkable behavioural patterns that distinguish the great apes from monkeys (Yamagiwa 1992; Idani 1995). Table 4.1 shows that those who peer are younger than those who are peered upon most of the time (89 per cent of 852 cases), which seems to indicate that the youngsters learn something from peering. Most peering occurs in the context of grooming (42.3 per cent of 852 cases) or feeding (42.7 per cent).[1] This finding highlights the importance of information on ectoparasites and food. That older individuals also occasionally peer at younger ones may mean that older chimpanzees must sometimes learn from younger individuals or refresh previously learned information.

Chimpanzees show great interest in ectoparasites such as lice, fleas, ticks, and mites. Leaf-grooming also attracts the attention of nearby chimpanzees, probably because the leaf-groomer is manipulating these ectoparasites (Zamma 2002).

Equally remarkably, chimpanzees show interest in the fresh wounds of their fellow group members. Peering occurs in the context of inspecting the wounds of others on 2.9 per cent of occasions.

Table 4.1 *Peering interactions at Mahale, 1999–2004* (N=852)

	Peerer younger than peered-at	
Behaviour in which peered-at is engaged	Frequency	%
Feed	364	42.7
Groom	360	42.3
Inspect wound	34	4.0
Total	758	89.0

[1] Nishida (unpublished data). Those who peered were younger than those that were peered at 89.0 per cent=758/852). Most peering occurred in the context of grooming 42.3 per cent=360/852) or feeding 42.7 per cent=364/852).

Fig. 4.2 Many chimpanzees gather around an old female, *Fatuma*, who picks out sand fleas.

Although this rate accounts for only a small proportion of all peering, this is misleading, as the opportunity to observe injuries occurs much less often than the chance to observe grooming or feeding. Chimpanzees not only check the wound, but also groom and lick the injured part carefully. Even small infants appear concerned about the wounded individual and lick the afflicted part if given the chance. I am not sure whether or not they sympathise with the patient, but their attitude alone may be comforting to the wounded comrade.

Consequently, ectoparasites and wounding, two health problems that are important in their normal daily lives, cause chimpanzees to congregate in a single place and to peer at the victims. As many as 11 chimpanzees gathered around a single elder female, *Fatuma*, who was sucking out sand fleas (Fig. 4.2).

4.2 PLAY

Humans seem to live their lives in order to play. Johan Huizinga suggested that play, ranging in its forms from combat, games, gambling, sports, and war, to law, poetry, philosophy, art, science, and politics, appears to be the foundation of human culture (Huizinga 1950). Play in wild chimpanzees, however, has not been extensively studied. Given its importance to human beings, this neglect in 'chimpanzee science' is surprising. Many behavioural biologists have mostly been interested in topics directly related to reproductive success, since the formulation in the 1960s and 1970s of kin selection as well as reciprocal altruism and its successful application to animal behaviour (Hamilton 1964; Trivers 1972, 1974; Wilson 1975; Dawkins 1976). This focus diverted primate ethnologists' attention from play because the function of play was not well understood. Few researchers have studied the play of chimpanzees in the wild in any detail.[2]

In the nineteenth century, Karl Groos (1898) proposed the hypothesis that play in animals is the act of practising the serious behaviours to be performed later in adulthood. One-hundred-and-fifty years later, this theory appears to hold true for many types of play.

However, although the practise theory of play may explain the case of immature individuals, it obviously does not account for play by adults. How do we explain the function of adult play? I believe that since play is important to the physical, psychological, and social development of immature individuals, it became an independent instinct in its own right. Moreover, play evolved to be accompanied by pleasure, just as sex, eating, and suckling – all critical for reproduction – are pleasurable. Consequently, play began to be used in communication among adults, just as feeding and sex are communicative in human beings. Similarly, sex is used for communication or social cementing among bonobos (Kano 1992; de Waal 1995; de Waal & Lanting 1997). Pleasure facilitates the most important activities for reproduction. Therefore, the fact that adults also play is not a devastating counterargument to the practise theory.

Recently, new interest in animal play has emerged among ethnologists, although dogs, cats, antelopes, and non-ape primates are still the major subjects of study (Bekoff & Byers 1998; Power 2000; Pellegrini & Smith 2005; Burghardt 2006). Play should be one of the major targets of chimpanzee studies: chimpanzees spend much time in play, and time spent in play correlates with cognitive abilities (Fagen 1981). In addition to the great apes, macaques, capuchin monkeys, dolphins, and such birds as crows and parrots, being famous among animals for being clever, they also play a lot (Diamond & Bond 1999)!

Here, I do not define play strictly, since such investigations are numerous (Bekoff & Byers 1998; Power 2000; Pellegrini & Smith 2005; Burghardt 2006), and one can easily distinguish it from other behaviour. Instead, I focus on behaviour that is performed repeatedly, without any immediate benefit in the related context, and which is likely to be accompanied by a play-face or play-panting. Chimpanzees play as

[2] Goodall (1968) and Hayaki (1985) were two exceptions.

much as humans. Although humans are often said to be the only animal that plays even after they mature, this is patently false, as even crows and cats play as adults (Fagen 1981). Play has been divided into solitary and social play. Solitary play consists of locomotor-rotational play and object play. Wrestling and chasing conspecifics are usually classified as social play, but these activities sometimes involve objects and locomotor-rotational elements. Therefore, solitary and social play cannot be divided strictly in chimpanzees. Nevertheless, here I treat patterns of locomotor-rotational play and object play as solitary play.

Locomotor-rotational play is useful for developing the ability to cope with and to avoid predators, to demonstrate fighting ability to conspecifics, and to feed in precarious settings. Social play is useful for promoting fighting ability against conspecifics and predators and for getting information on the fighting abilities and personalities of others.

4.3 LOCOMOTOR-ROTATIONAL PLAY

4.3.1 Early development of locomotor-rotational play

Chimpanzees younger than six months old do not engage in social play. Instead, they 'climb' their mothers, and as they grow, they walk a bit on the ground, climb short shrubs, and hang from branches with one or both hands. One-year-old and older infants hang from a branch by one hand and spin. They cling to a woody stem or vine and shake it back and forth with hands and feet. They swing forward and rotate upward like human children hanging onto a horizontal bar. In the middle of the second year of life, aged around 17–18 months, chimpanzees begin to perform acrobatics such as somersaults, pirouettes, and forward swinging, but their performance is clumsy.

An immature chimpanzee on its own sometimes runs around and around a tree while emitting play-pants. It may run around the tree bipedally while holding a stem with one hand ('circle orthograde') or on all fours ('circle quadrupedal') (Nishida *et al.* 2010). This type of play, which also begins at roughly one year of age, is more often performed as social chasing (see below).

4.3.2 Somersault and pirouette

Rolling head over heels produces forward and backward somersaults. Side somersaulting also occurs. Forward somersaulting is most frequent.

Older infants (≥2.5 years) and young juveniles often somersault forward while following their mothers during long-distance travel. They may repeat this more than five times in a row without any direct benefit. As with human children, sandy soil stimulates chimpanzee youngsters to play. I have often seen an infant rolling, somersaulting, or sliding over sandy ground. A tourism company once put sand piles at the crossings of main observation paths to install guideposts. The piles were conical in shape and about 0.8 m high and 2 m in diameter at the base. A juvenile female, *Acadia*, rolled and somersaulted alone on a sand pile while emitting play-pants.

During travel, a young chimpanzee also pirouettes or spins around in a quadruped posture while advancing forward in a straight line (Goodall 1968; Nishida & Inaba 2009). A seven-year-old juvenile male, *Xmas*, performed the maximum number of 14.5 cycles, and an eight-year-old juvenile male, *Cadmus*, attained the maximum speed of 1.83 cycles per second (Fig. 4.3).

Immature chimpanzees seem to make every effort to learn to perform somersaults and pirouettes quickly and repeatedly. They apparently strive to improve their skills in the same way human athletes do. The effort is spontaneous and deliberate: they are not encouraged to do this by their mother or any other individual. Remarkably, we sometimes get the impression that young chimpanzees display pirouettes to

Fig. 4.3 A juvenile male, *Xmas*, pirouettes.

conspecifics and even human observers. The arrival or departure of a particular individual, such as a mother, sister, playmate, or adult male sometimes seems to elicit pirouettes. Youngsters watch these members of their audience while they are pirouetting. Moreover, chimpanzees also performed pirouettes immediately after they displayed to me, after slapping the ground, charging at me, or throwing a stone at me. My 'target' chimpanzee also pirouetted after gently touching my trousers or after I retreated to an appropriate distance of 5 m from an individual who had moved away.

Our observations thus clarified that chimpanzees seem to be interested in attracting the attention of others. I have speculated on the origins of human spectator games like the Olympics. Why are humans attracted so much to performances by other persons? Human performers likely show-off for the crowd, and this may be the common root in locomotor-rotational play patterns between humans and chimpanzees.

Both species are interested in attracting others' attention. Displaying superb athletic capability may be useful for impressing others – for example, impressing opposite-sex spectators for acquisition of future sex partners, or older and younger competitors for acquisition of social status.

4.3.3 Elaborate patterns of locomotor-rotational play using objects

Older infants or juveniles walk quickly backwards while raking dry leaves with both hands (Fig. 4.4a, b). They seem to enjoy the noise caused by the raking and accumulation of dry leaves, and when they gather enough leaves they stop moving and sometimes hug all of the leaves and somersault sideways while scattering them little by little. We call this 'leaf-pile pulling' (Nishida & Wallauer 2003). No such play pattern had been reported elsewhere before, although this is now known to occur at Gombe too, but much less frequently. This play pattern seemed to emerge at Mahale in the late 1980s, because we did not see it before then. However, since it is now done by almost all immature chimpanzees in M-group, we can call it an M-group custom. The prevalence of this innovative pattern may have been facilitated by our efforts to maintain good observation paths, such as cutting down small shrubs that could hinder walking backwards. In 2001 I had a key observation regarding social learning: a ten-year-old male, *Orion*, was carrying his three-year-old brother, *Oscar*, on his

(a)

(b)

Fig. 4.4a, b *Primus* rakes dry leaves (leaf-pile pulling).

back (Nishida & Wallauer 2003). When *Orion* came to a slope, he turned around and began to engage in leaf-pile pulling, which continued for 15 seconds, while their mother, *Opal*, walked after them. After *Orion* ceased the pattern and let *Oscar* down, *Orion* walked on ahead. Within one minute, *Oscar* turned around and engaged in leaf-pile pulling for 8 seconds while walking immediately ahead of his mother (Fig. 4.5a, b). It is doubtful that the 'imitation' occurred only at

Fig. 4.5a, b *Orion* and then his young brother, *Oscar*, leaf-pile pull in front of their mother, *Opal* (drawing by Agumi Inaba, from video clips).

this time, since such opportunities for social learning are probably plentiful in daily life (see Chapter 9 for more on imitation). Video footage taken by Miho Nakamura in 1989 shows that an adolescent male, *Hanby*, performed leaf-pile pulling immediately after an adolescent female, *Penelope*, showed the same behaviour. This footage shows that social facilitation (*sensu* Thorpe 1963) occurs frequently in the lives of wild chimpanzees. Therefore, the observations hint to us that a custom can be transmitted from older to younger brother or via playmates.

Leaf-pile pushing (Nishida & Wallauer 2003) is less common, but is also shown by many youngsters of M-group. On all fours, they push dry leaves with both hands forwards, and produce rustling noises as well. Interestingly, leaf-pile pushing is an element incorporated occasionally in charging displays by some adult males, such as *Fanana* and *Masudi*. Consequently, this behaviour may provide evidence for the practise theory of play. It is likely that this is a tradition maintained by adult males. Leaf-pile pulling, however, has never been observed during the charging displays of adult males, presumably because this is a backwards movement. It is a performance shown only by immature chimpanzees.

4.4 OBJECT PLAY

4.4.1 Objects used

Infants in the first year of life may take a dry leaf or tiny piece of vegetation such as a piece of fruit skin from the ground and hold it between their lips. This behaviour of picking up an object by hand and then holding it in the mouth is the first step of object manipulation by chimpanzees. A one-year-old may pick up and hold in the mouth larger objects such as a branch, vine, bark, or stone. Youngsters touch, lick, beat, manipulate, carry, toss, or throw various objects found in the environment: dead or live branches, vine segments, dry or green leaves, fruit, fruit rinds, petals of flowers, insects, bones, pieces of animal skin, bird feathers, whole bird carcasses, sand, stones, water, moss, or artefacts such as pieces of cloth, paper, and discarded earthenware. They carry them by hand, in the mouth, in the neck pocket, in the groin pocket, or on the back or head. They discard these objects when their curiosity seems satisfied.

One- to five-year-old infants sit on the ground, grasp and lift up sand with one hand, and let it fall little by little from their upraised hand. When all the sand has fallen away, they may repeat the same action. This can happen several times. They also rake sand with both hands on the ground. A chimpanzee may begin to carry a stone at one year of age and lift it high overhead at two years. Throwing begins with scattering dry leaves bipedally at age two, and then throwing small

Fig. 4.6 *Michio* rolls *Myrianthus* fruit with his hands and feet.

branches bipedally later. Only after three years of age do infants begin throwing stones, but only rarely. Some six-year-old males attain the ability to skilfully throw a stone with an overhand motion.

Once, a six-year-old male lay on his back and used both hands to roll a softball-sized fruit placed on his feet (Fig. 4.6), just as a dolphin rolls a ball with its snout. This ball-rolling was shown by another female at a different time.

4.4.2 Leaf sponge and leaf spoon

The chimpanzees of Gombe insert leaves into the mouth, making a sponge-like device, dip it into trapped water in a tree's hollow, and drink (Goodall 1968). This feat is so famous that it is included in a Japanese school textbook. But I never saw it done at Mahale before the 2000s. In 1987 Hiroyuki Takasaki and his assistants saw two females from K-group perform this 'leaf sponging' pattern only a few times. It was not yet part of M-group's repertoire (Nishida 1990b).

Since the early 1990s, we have noticed that chimpanzee youngsters very often play in streams. Takahisa Matsusaka and colleagues were the first to see the 'leaf sponge' and 'leaf spoon' patterns (Matsusaka *et al.* 2006). The former is the same as that recorded for Gombe (Fig. 4.7a, b). Youngsters of Mahale also use a leaf spoon: they pick up a single large leaf, dip it into a stream, and lap up the water in the 'spoon'. One of the juvenile males first leaf-clipped and dipped the

(a)

(b)

Fig. 4.7a, b The leaf sponge.

midrib section into water (Fig. 4.8a, b). Of course, if they were really thirsty they would drink directly from the water's surface rather than sipping it from a 'leaf spoon'. Therefore, this is a type of play, not a subsistence activity. They use leaf sponges in both streams and in tree hollows.

(a)

(b)

Fig. 4.8a, b The leaf spoon.

4.4.3 Water play and self-image in water

Youngsters of M-group put their fingers or a hand into the water of a stream and stir or splash water. They may enter the stream and push or pull rocks. More often, they look intently at the watery surface of a stream. An adolescent male, *Cadmus*, an adult female, *Totzy*, and a juvenile male, *Michio*, shook their heads up and down and watched the

(a)

(b)

Fig. 4.9a, b An adolescent male, *Cadmus*, vigorously nods his head up and down while watching movement of his reflection on the water's surface.

movement of the reflection of their faces on the water's surface (Fig. 4.9a, b). They often play-panted while doing this (Nishida *et al.* 2009). Captive chimpanzees are known to recognise themselves in a mirror image. When a chimpanzee wakes up after being anaesthetised and having had a dye-spot applied to the face, it often immediately touches the marked part of the face while looking at a mirror (Gallup 1970).

Some chimpanzees in the natural habitat may recognise their self-image in the reflection of a stream. At least 14 M-group chimpanzees (1 adult female, 13 immatures) were observed to gaze intently at the water's surface of streams. Moreover, six of them nodded their heads vigorously while watching their reflections. An adolescent male, *Cadmus*, not only watched the reflection of his movement, but also showed play-faces while nodding his head vigorously. I do not assert that this is evidence of mirror self-recognition, but it is the best observation available to suggest it in the natural habitat.

4.5 SOCIAL PLAY

As described above, chimpanzees younger than six months do not engage in social play. Until the first half of the second year of life, infants rarely make contact with individuals other than their mothers. One-year-old chimpanzees first play 'in parallel', namely performing solo play in close proximity to other peers. They also begin to chase and 'wrestle' with each other, albeit briefly and clumsily. Older infants and juveniles are major players. Two individuals typically play, and occasionally 1–2 others may join them. However, as time passes play returns to a two-person game, in which typically only the original two players remain; one of the original players is rarely displaced by a new participant. An individual who takes over the play is inevitably older than at least one of the original players.

4.5.1 Following, or chase-and-flee patterns

A one-year-old infant may climb a shrub or tree and climb or leap down, while an age-mate follows suit. They may repeat this 'climb-and-descend' game almost endlessly. They rarely make physical contact, and so this may be regarded as parallel play, or simply as locomotor-rotational play.

Simple chase-and-flee games rarely occur among infants. A juvenile or adolescent might chase an infant in a straightforward manner. However, among infants and juveniles, chasing and fleeing games take the form of chase-and-turn-around play.

4.5.2 Chase-and-turn-around

Probably the most popular play among generations of M-group chimpanzees, shown by both immature and mature individuals, is chase-and-turn-around, in which older infants are the major players.

(a)

(b)

Fig. 4.10a, b Circling: juveniles circumnavigate a tree while playing.

One partner chases or follows another around a tree or another individual, or without any central focal-point, but just around and around, making a big circle many times (Fig. 4.10a, b). A chased or followed individual sometimes approaches so close behind the chaser or follower that we cannot judge who is the chaser or chased. Or, the chased sometimes reverses direction to take the role of chaser. Such role reversal has often been the characteristic of social play (Fagen 1981; Bekoff & Byers 1998; Power 2000; Pellegrini & Smith 2005; Burghardt 2006).

When the chaser catches the chased, they wrestle before resuming their running around. They may repeat this sequence several times. While they chase-and-turn-around, either or both of the players launch into forward somersaults, often at a fixed point of the circle. Youngsters often select a 'tunnel' in the undergrowth and pass through it in the course of the cycle, presumably in order to increase enjoyment. As with solitary play, sandy soil attracts youngsters and stimulates them to engage in chase, circle, and somersault play. Juveniles often begin chase play when a big party in which they are moving passes the sandy ground beside a small stream. Chase-and-turn-around is also seen at Gombe (Goodall 1968) and among the bonobos of Wamba (Enomoto 1997).[3]

4.5.3 Wrestle, hang-wrestle, and finger-wrestle

Wrestling is immensely popular among older infants, juveniles, and younger adolescents, and is certainly the most popular type of social play for juveniles (Fig. 4.11). The aim of wrestling seems to be mouthing the partner on any body part, as Aldis pointed out long ago (Aldis 1975). A player makes every effort to mouth the other player while, on the other hand, avoiding by all means getting mouthed by the play partner. This two-sided endeavour is the wellspring of all behavioural patterns involved in wrestling, such as push, pull, kick, bite, and hold. The mouthed player, rather than the mouthing one, often play-pants, the function of which is assumed to maintain the interaction (Matsusaka 2004).

Adults also wrestle with infants and juveniles, but mainly they tickle or mouth youngsters, rather than vice versa. Young males wrestle more often than their female counterparts.[4]

One-year-old infants begin to 'hang-wrestle' with another (often older) infant while suspended above the ground. Both hang from a

[3] When I watched the video shown by Isabel Behncke in the Congress of IPS 2010, I noticed *juvenile* bonobos seem to engage in this type of play more often than *infant* bonobos and without adding somersaults.

[4] Goodall (1968) and Hayaki (1985) were two exceptions.

Fig. 4.11 Two adolescent females play-wrestle on the ground.

horizontal branch and reach an arm to each other and pull, push, or mouth the playmate. This often results in one wrestler falling to the ground. An older infant or juvenile often tries to detach the hand of the playmate from the branch. This is a deliberate tactic to make the playmate fall. The loser, however, returns quickly to a play-fighting position and the hang-wrestling continues.

In another form of wrestling, a chimpanzee lies supine and stretches its hand to another's body part, such as a hand, foot, or armpit, and begins to tickle the playmate on the ground. If the victim grasps the attacker's hand or avoids being tickled, or even tries to tickle the other's body parts, this is a gentle version of wrestling, which Goodall (1968) termed 'finger wrestling', which might sometimes be better called 'hand-wrestling'. This type of social play is almost always confined to adults or older adolescent chimpanzees (Fig. 4.12). If it becomes rougher, they begin to play-pant and may even pummel each other.

4.5.4 Mother-offspring play

Mothers and infant or juvenile offspring often play together. When the offspring is a tiny infant less than one year old, the mother takes the initiative in play and tickles or mouths the infant one-sidedly. When the offspring grows older, they may wrestle and play

Fig. 4.12 An old female, *Gwekulo*, and an adult male, *Fanana*, finger-wrestle.

Fig. 4.13 *Xtina* and her juvenile son, *Xmas*, wrestle.

chase-and-turn-around. This is the case for either a mother–son or a mother–daughter pair. Either or both of them play-pant and show play-faces (Fig. 4.13). Mother–offspring play is gradually replaced by peer play, but they continue to play until the offspring reaches around eight years of age, the end of juvenility.

4.5.5 Social play between infants and adolescents or nulliparous adult females

Adult females with infant offspring rarely play with other infants (Nishida 1983b), although sterile females may do so. Mature adult females without their own infants occasionally engage in such play, and adolescent females and young nulliparous adult females play with infants most often (Fig. 4.14). However, their play is mostly a form of 'play-mothering' (Lancaster 1972) or babysitting. Although nulliparous females seem to prefer younger to older infants as the object of babysitting, the mothers of infants usually consent to handing over only their older infants. Thus, whether or not an infant is transferred to another's care is negotiated between babysitter candidate and mother. Whether a younger infant (<1.5 years) is put in the care of a nulliparous female depends upon how regularly and how long she has groomed and associated with the mother. If the nulliparous female has recently immigrated from another group, she begins by continuously following a particular high-ranking, middle-aged female, beginning the contact by grooming her one-sidedly, and then caring for her infant. The relationship formed is likely to give benefits to both parties: the labour saved in carrying for and supervising the infant for the mother, and practice in infant-handling and self-protection gained by the

Fig. 4.14 An unrelated adolescent female, *Wakilufya*, transports *Chausiku*'s infant, *Katavi*.

nulliparous female (see Chapter 5). Although no such relationship has been reported from other chimpanzee study sites, a nulliparous immigrant female bonobo forms a similar relationship with a resident female (Idani 1991).

A chimpanzee mother or allomother returns to her infant charge when the infant is stranded in a gap between trees or between trees and rocks. She seems to recognise the plight of the infant. I have never seen mothers fail to respond to infants stranded in such a gap, except during the weaning period. This rescuing response appears to be very primordial and is not surprising to us. However, this does not seem so obvious when we consider that baboon mothers do not respond with such care. Baboons do not return to a river crossing point, even when their juvenile offspring cannot cross the flooded river and calls for help. Thus, many baboon juveniles are killed by predators or are drowned. Baboons do not seem to understand the plight of their offspring – that is, they seem unable to see from others' perspectives (Cheney & Seyfarth 2007).

A chimpanzee allomother anxiously returns her infant charge to the mother as soon as it begins to whimper for suckling milk. A babysitting adolescent female sometimes even throws a temper tantrum if she cannot immediately find the charge's mother. She understands that she cannot satisfy the infant's demand by suckling it.

A mother never hands her infant to others with whom she has had scarcely any social interactions such as grooming (Nishida 1983b). She delivers her infant only to reliable babysitters. They are either her elder offspring, a new young immigrant who has associated with the mother, or a particular infertile resident female, such as *Gwekulo*, who has cared appropriately for many infants.

4.5.6 Social play between adult males and infants

An adult male, usually an older rather than younger individual, may approach an infant with a play-face and tickle or mouth it. If the infant flees, the adult male chases it and engages in a chase-and-turn-around game. An adult male may grasp the hand of an infant in his mouth and drag the infant. He may lie supine and lift the infant up and down with his feet, thus playing 'aeroplane'. The mother usually monitors the development of such social play intently (Fig. 4.15), and if the infant is young and the male's actions become too rough, she may approach him with pant-grunts and try to retrieve the infant. However, great care

Fig. 4.15 A mother monitors ongoing interactions of infant son and adult male alloparent.

by the mother is necessary, because such an intervention may provoke the male to show aggression against the mother.

4.5.7 Social play in adults

Adult males and females occasionally engage in finger-wrestling, but less often they engage in circling or chase-and-turn-around. Those who show such social play are middle-aged adults with a long history of co-residence. For example, *Kalunde* was often seen to play finger-wrestling or chase-and-turn-around with *Gwekulo* or *Nkombo*, his two long-term girlfriends. *Kalunde* and these females spent more than 20 years together in M-group.

Adult males play only with their friends and potential coalition partners. For example, when *Fanana* dethroned *Kalunde* and became alpha, and *Kalunde* was threatened by younger males such as *Alofu*, *Kalunde* depended on *Fanana* for agonistic aid. *Fanana* and *Kalunde* became strong coalition partners. At that time, they often rested together, reciprocating grooming and playing finger-wrestling.

Among adult males, play is also a means of reconciliation. When *Ntologi* returned to M-group after an eight-month, semi-solitary life following his banishment by *Kalunde* and *Nsaba* (see Chapter 8), they accepted him by pant-grunting. Kazuhiko Hosaka observed *Kalunde* and *Ntologi* embracing and giving each other an open-mouth kiss. They also played by engaging in chase-and-turn-around, mixing it with prolonged wrestling. Thus, the *Ntologi–Kalunde* coalition was re-established and *Nsaba* was robbed of the opportunity to become alpha for three more years. Accordingly, social play among adult male chimpanzees seems to be an important way to cultivate and maintain friendships and alliances.

4.5.8 Self-handicapping

An older chimpanzee often lies prone and allows a younger playmate to climb onto his belly and mouth him. An adult male could easily pin down the infant from the beginning to the end of such play. This behaviour cannot be simply play, as the male does not assert his overwhelmingly dominant status over his subordinate playmate. He could easily send his playmate flying, but instead continues to assume the subordinate posture. This type of behaviour is called self-handicapping. Furthermore, in the course of playing, chimpanzees may change their roles: the dominant male may chase the subordinate playmate and then, suddenly, the subordinate looks back, reverses direction, and begins to chase the dominant. The dominant individual apparently accepts this role change. Self-handicapping and role-changing are two important ways used to engage in social play such as wrestling.

4.5.9 Solicitation of play: gestural and postural communication

Unlike object play and locomotor-rotational play, social play needs the cooperation of a partner. The player must not only monitor the mood of the partner, but also inform him or her of the intention of play – namely, engage in meta-communication (Bateson 1955). A chimpanzee has various ways of inviting partners to play. If the intended partner is the mother, a son or daughter may touch her with or without play-face. She immediately understands the desire of her offspring. If the invited chimpanzee is a frequent playmate, it is enough to touch, push, or lightly kick him or her with a play-face or simply to stand and gaze at the intended partner. These signals develop in step with 'ontogenetic ritualisation' (see Chapter 9). Here, I discuss several key aspects of chimpanzee play and how chimpanzees convey their desire to play.

Fig. 4.16 *Cynthia* invites her infant son, *Caesar*, by looking back through her legs.

First, there are special postures and gestures that characterise play. By showing an unusual posture or gesture that is not performed in a serious situation, a chimpanzee implies that this is play. This may be called a visual attention-getter. Juveniles may 'play walk', 'tilt head' or 'nod' with 'play-face'. Holding an object in the mouth is a common pattern of chimpanzees, referred to as 'play start' (Whiten 1999). A mother, *Cynthia*, invited her infant son, *Caesar*, by standing on all fours with her buttocks directed to him and looking back at him through her thighs (Fig. 4.16), and a juvenile female assumed the same posture when inviting an adolescent male (Fig. 4.17).

Second, patterns of solo play can be useful as indicators of the desire for social play. Thus, a chimpanzee may somersault, circle, swing, or hang in a sloth position while looking at a potential play partner.

Third, postures, gestures, and movements that are important elements of social play, in particular wrestling and chase-and-flee play, can be used to invite play. Play may begin with a flail, grab, hit, pull, push, or 'hit and run', plus play-face. Youngsters may simply start to flee while looking back at a potential play partner. All of these patterns are intention movements that are likely to be used as communicatory signals (Tinbergen 1951). Similar gestures such as shaking an arm are employed when youngsters play above the ground. Adult males use tickling to initiate play with infants.

Fig. 4.17 *Maggie* invites an adolescent male, *Jilba*, to play by looking back through her legs.

Fourth, self-handicapping gestures and postures may be used to solicit play. If a juvenile or older individual wants to play with an infant, the instigator may lie supine (or, more rarely, prone) on the ground to invite the infant to ride on him or her. The reclining posture implies a relaxed mood and the inability to mount an instant attack, thus lowering the guard of the infant and its mother. 'Gazing' at the desired playmate while lying down is necessary in this case.

One day, during the afternoon rest period, the alpha male, *Alofu*, approached a middle-aged, low-ranking male, *Masudi*, and suddenly performed a headstand and play-face in front of him. Later, they began a chase-and-turn-around game, with both of them play-panting a lot. The precarious posture of a headstand significantly weakened the attacking capability of *Alofu*, thus symbolising the lack of any intention to attack *Masudi*.

Fifth, we should consider other gestures classified as 'audio-visual attention-getters'. These include bending shrubs, slapping the ground, stamping on the ground, shaking branches, and slapping of the belly. These play signals are also used in other contexts, such as intimidation and courtship.

In solicitation of play, many gestures and postures are combined in turn rather than performed singly. For example, a seven-year-old juvenile female plucked a branch and then showed a headstand with the branch in her neck pocket, followed by somersaulting forward and lying in a supine position. After this, she stood quadrupedally, looking behind through her thighs. Consequently, her infant playmate came to her. In another case, an eight-year-old male stood quadrupedally, nodding repeatedly with play-face, while watching an adolescent playmate. Then, he flailed his arm and stamped on the ground.

Whether these play solicitation signals are particular customs of Mahale is unknown. First, some gestures might be unique to a few individuals of M-group, but not common to all individuals. Second, no detailed descriptions of play solicitation at other study sites have been published, except for Gombe.

Gestures such as leaf-clip and shrub-bend are also used in the context of courtship (see Chapter 5). Sitting-hunch, bipedal swagger, opening thighs, and branch-shaking are four courtship gestures common to the species, along with showing penile erection. Presenting is a well-known universal posture of oestrous females. In addition, beg, peer, touch mouth, and so on are important gestures used in social interactions involving food transfer. However, an overwhelmingly larger number of gestures are used in the context of play solicitation compared with other contexts (Nishida *et al.* 2010). This also holds among captive immature chimpanzees (Call & Tomasello 2007) and bonobos (Pika 2007).

Why are there so many signals in play? I assume that this is so because play and exploration are fundamental means by which immature individuals explore the surrounding world, including their conspecifics. Immature individuals try out possible ways to manipulate the world around them. Variable and flexible patterns are both adaptive to individuals and bring them pleasure. The implication is that play is the kernel giving birth to physical and social tactics in chimpanzee life. During evolution, play became an independent instinct or drive due to the important role it played in physical and social development.

It is possible that ancestral humans began to use play solicitation gestures for invitation, request, and order in various other contexts. This might be related to the origin of human speech from gestural communications.

4.6 COMPARISONS WITH HUMAN PLAY

Compared with play among non-human primates, including chimpanzees, there are two conspicuous characteristics of human play (Nishida & Inaba 2009). One is the presence of onlookers, who do not engage in the play activities but just watch them, although this does not always occur. The other characteristic is group play (Nishida 2004), in which one team of players competes against another team of players.

4.6.1 Presence of onlookers

The presence of spectators seems to be related to the size of the society or community. More members means more onlookers. However, so far as I have observed, even in a large provisioned troop of macaques, there seem to be no onlookers of play activities. M-group once had more than 100 chimpanzees. Even at that time, we saw no spectators of play. Mothers occasionally monitor the development of their offspring's social play, with adult males in particular. However, this simply shows the mother's concern for her offspring, not behaviour like that of a spectator at a baseball game. Youngsters also watch fellow chimpanzees' play interactions. But they are not simply spectators, but instead on the lookout for an opportunity to join in the play. Of course, humans also play without spectators, but they are still the only animal species that enjoys play as spectators. Why is this so?

Humans seem to be one of the most empathetic animals on Earth. When I watch Japanese sumo wrestling on television, I unintentionally clench my fists. I feel compelled to push hard or throw down my rival with all of my strength. Subconsciously, I seem to identify with one of the sumo wrestlers. I notice that I even cry out as soon as I have 'won' or am 'defeated'. Simply put, I am vicariously experiencing the sumo match. I am sure that all normal humans have felt such an emotion at one time or another. The strong traits of human curiosity and imitative ability may be related to the phenomenon of spectators at events of other people's play.

4.6.2 Lack of group play

Human group play may be one of the cultural universals. Why is this so? The function of animal play behaviour has been discussed by many authors. As mentioned above, Groos proposed the hypothesis that play is practice for serious behaviour that will be of use in the future. For example, play-fighting may be practice for fighting among conspecifics or defence against predators. By studying play-fighting among mammals, and macaques in particular, Aldis (1975) and Symons (1978) provided persuasive evidence to support the practise hypothesis.

Play-fighting among humans is exceptional in that they engage not only in individual-to-individual play, but also group play. By *group play* I mean that two groups of individuals compete or mock-fight for victory. People are excited to engage in, or observe, tug-of-war (Turnbull 1965), baseball, football (Morris 1981), rugby, basketball, volleyball, rowing, etc.

Even in card games, you can find 'Napoleon' or 'bridge', where two coalitions 'fight' against each other. If region-specific games are included, we can find many more examples. For example, two groups of Yanomamo boys play-fight with long sticks that can injure a player badly (Chagnon 1992). As another local version of group play, in Japanese schools, 120 boys are divided into two groups of 60. Each group consists of 15 quartets, where a quartet is made up of three boys who form a 'warfare cart' and the fourth boy who rides on them. The boys riding atop the carts 'fight' with extended arms against any rider of the other team. When either removes the cap of the rival, this is considered a win; the victorious team is the one with the greatest total numbers of snatched caps. Performing some kind of group play may be one of the human universals, although the most thorough review of human universals does not include it as such (Brown 1991).

On the other hand, I have never heard of social animals splitting into two groups and competing or play-fighting against each other for victory. Even the most conclusive reviews of animal play published so far contain no examples of group play in animals (Fagen 1981; Bekoff & Byers 1998; Power 2000; Pellegrini & Smith 2005; Burghardt 2006). If the practise hypothesis is valid, that play's function is the practise of behaviour that will be useful in the future, the function of group play in human beings would appear to be practise for warfare, or an organised battle between two groups.

In the animal kingdom, coalitions are mostly formed against a single individual (Harcourt & de Waal 1992). In within-group contests among chimpanzees, a contestant assisted by a third party fights against his rival (de Waal 1982; Nishida 1983a). Gang attack is always directed to only one or at most two individuals. An exile, or ostracised ex-alpha male, may be chased fiercely (Nishida 1994) or severely attacked (Goodall 1992) by a group of chimpanzees. In one observation, a young adult male who had not greeted his superiors was severely attacked by the alpha male and his seven coalition partners (see Chapter 8). In a possible example of sexual competition, a young, low-ranking adult male was killed by many adult males of the same group (Fawcett & Muhumuza 2000). Even in antagonistic confrontations between unit-groups of chimpanzees, only one side is an 'organised' multi-male party, with the counterpart usually being a lone individual who is victimised (Goodall 1986; Nishida 1979; Muller 2002). Therefore, Boehm's (1999) 'macro-coalition', the terminology coined for between-group conflicts, might be a misnomer for chimpanzees. An exception may be the case of male dolphins: two or three male dolphins unite forces against another similar coalition in competition for fertile females (Connor *et al.* 1992). If my theory is correct, dolphins may be the only animals other than human beings in which group play can be observed.

Perhaps Konrad Lorenz (1966) stated a concept that parallels my theory when he suggested that sports provide a good outlet for human aggressive impulses that otherwise seek warfare. However, he did not remark explicitly that no other animal engaged in group play, since at that time there was very little detailed study on large-brained animals such as chimpanzees and dolphins.

If my argument is correct, bellicose tribes or nations may encourage group play among the youngsters of their kind. This notion could be corroborated or refuted by a comparison of ethnographic data on tribal warfare.

5

Communication as culture

Chimpanzees communicate with conspecifics by facial expression, posture, gesture, locomotion, vocalisation, and odour. Here, I do not dwell on detailed description of various types of communication. Interested readers will find useful definitions in various published books and papers, as well as some relevant descriptions in each chapter of this book.

Local differences in tool-use among chimpanzee populations are well-known (see Chapter 2). Here I highlight local differences in patterns of social behaviour, the remarkable differences in communication from subtle gestures to more obvious patterns of courtship and vocal communication. Studies on the chimpanzees of Mahale have contributed much to the development of research on local differences of social behaviour.

5.1 SOCIAL GROOMING

Chimpanzees spend 4–23 per cent of daytime hours in grooming.[1] They often form a large cluster of social groomers, which Michio Nakamura has called a 'clique' (Nakamura 2003).There are many types of clique: for example, straight-line grooming, in which A grooms B, B grooms C, and C grooms D, thus forming a straight line; or circular grooming, in which the configuration above is completed if D grooms A. A and B may groom each other mutually, and C may groom B. Nakamura found 25 types of clique, and the largest one comprised 15 individuals. Why do so many chimpanzees congregate to groom? Nakamura considers that they save time by simultaneous grooming, as they have many social commitments, such as to family, sexuality, politics, friendship, etc. By contrast, monkeys huddle to form even bigger clumps in cold weather

[1] Wrangham (1977): 4.1–8.8 per cent; Nishida (1981): 9.0–23.3 per cent; Nishida (1989): 5.4–8.3 per cent.

(Anderson & McGrew 1984), but unlike chimpanzees they make only physical contact and do not groom each other.

Grooming removes ectoparasites such as lice, and so is beneficial to the groomee. The groomee reciprocates by giving grooming, sex, meat, aid in combat, opportunities to practise mothering, etc. Grooming becomes the currency of communication. It sustains and strengthens social relations, renews friendships in reunion, reconciles rivals after confrontations, and encourages companions to form coalitions against a common rival. That apes understand the value of social grooming is shown by the fact that an alpha male chimpanzee harasses grooming between his rivals by 'separating intervention' (see Chapter 8), or that an adolescent or juvenile female one-sidedly grooms a mother with a small infant in order to gain access to the infant for play-mothering (see Chapter 4).

5.1.1 Grooming hand-clasp

When Bill McGrew and Caroline Tutin visited Mahale in 1974, after having studied at Gombe, they noticed a strange (to them) style of mutual grooming at Mahale. (I was not at Mahale then.) When the chimpanzees of Mahale groom one another simultaneously, they often lift either the right or left arm straight up in the air, clasp hands, and then proceed to groom the underarm of the other arm that is not clasping (Fig. 5.1). They termed this perfectly symmetrical posture the

Fig. 5.1 The grooming hand-clasp.

Fig. 5.2 Grooming branch-clasp (courtesy of William McGrew).

'grooming hand-clasp' (hereafter, GHC). Interestingly, the chimpanzees of Gombe do not practise the GHC but instead only grasp an overhead branch (Fig. 5.2). When they published the seminal paper on chimpanzee culture (McGrew & Tutin 1978), I was surprised because this grooming gesture was one I never imagined would be absent at Gombe. I had no idea how this discrepancy arose. I speculated that this posture may have emerged to ensure that each party get a fair share of grooming. If so, then why did counterparts at other study sites not hit upon the same idea? Now, it is known that the GHC occurs at Kanyawara, Ngogo, Kalinzu, Semliki (all sites in Uganda), and at Lope (Gabon), and in a captive colony (Yerkes). Only some individuals at Taï (Ivory Coast) do it. However, it does not occur at Bossou (Guinea) and Budongo (Uganda), in addition to Gombe (Nakamura 2002). Presence or absence of GHC shows no relation to the geographical distances between sites. In bonobos, at least some group members of Wamba, Lomako, and Lui Kotale do GHC (Nishida *et al.* 2010).

It is interesting how this hand-clasping tradition has developed. When McGrew and Tutin first described the GHC, chimpanzees of K-group (M-group was not habituated well then) grasped hands with a palm-to-palm clasp. However, in the late 1990s, when McGrew and Linda Marchant observed the chimpanzees of M-group, they found that they irregularly engaged in GHC (McGrew *et al.* 2001); often palm-on-wrist, and even touching only the dorsal parts of the hands together. In the latter, abbreviated case, participants do not 'grasp' hands at all. So, K- and M-group seemed to have different customs of GHC. Nakamura and Shigeo Uehara extended the earlier findings in detail, on the basis of many more samples. It turned out that rather than palm-to-palm, 'straight elbows' may be K-group's norm, and 'flexed wrist' may be M-group's norm (Nakamura & Uehara 2004).

When an adult female, *Gwekulo*, was in K-group, she showed 'straight wrist' and 'straight elbow', thus yielding the palm-to-palm grasp. After she immigrated to M-group, she continued the straight wrist but compromised her elbow posture by flexion, although she still showed more often the straight elbow. What is interesting is the arm posture of *Gwekulo*'s close associates, *Pinky* and *Primus*. *Gwekulo* was a regular 'auntie' of *Primus* when he was 3–5 years old, and *Pinky* became *Gwekulo*'s close friend (see Chapter 6). *Primus* was the only straight-elbow groomer in M-group, and *Pinky* was for half of the time a straight-elbow groomer (Nakamura & Uehara 2004). It is likely that *Pinky*'s arm posture in GHC was influenced by *Gwekulo*. So how was *Primus*' arm posture shaped? I speculate that *Primus* learned GHC from his mother, not directly from *Gwekulo*, because I never saw *Primus* and *Gwekulo* grooming mutually.

Since 1999 I have studied innovative behavioural patterns and their propagation in M-group. I focused on observations of social interactions between mothers and immature offspring and among youngsters. A video-tape recorder was useful because, in addition to the audiovisual data, I could simultaneously narrate my observations.

In 1999, I saw *Pinky* lift the left hand of the ten-year-old *Primus* with her right hand, shift the gripping side from right to left hand, and groom his left armpit with her right hand. *Primus* responded to her by grooming her left armpit in an exact mirror image. It was a typical GHC. I began to look for another example of the GHC sequence in mother and younger offspring.

In the same year, while I was observing an adult female and her four-year-old son, she took his right arm with her left hand, lifted it upward, shifted her gripping side from left to right hand, and began to groom his right armpit with her left hand. The infant son did not groom his mother this time, since he was too young. In 2002, another female lifted the left upper arm of her six-year-old son with her right hand and then gripped his left wrist with her left hand to groom his left armpit. Like *Primus*, the son responded to her by grooming her left armpit in typical GHC manner.

I remembered that an ape-language psychologist sometimes 'shapes' a chimpanzee hand to form an appropriate manual sign when teaching human sign language. This is called 'moulding' (Gardner & Gardner 1969). The mothers' behaviour as described resembles this manual shaping. In particular, the mother had to take the initiative when the offspring was still an infant and not accustomed to social grooming, much less the GHC pattern. Mothers did not force their offspring to groom them mutually. However, the sons may have understood that they were expected to groom their mothers as the sequences were repeated (Fig. 5.3). Human moulding is intentional,

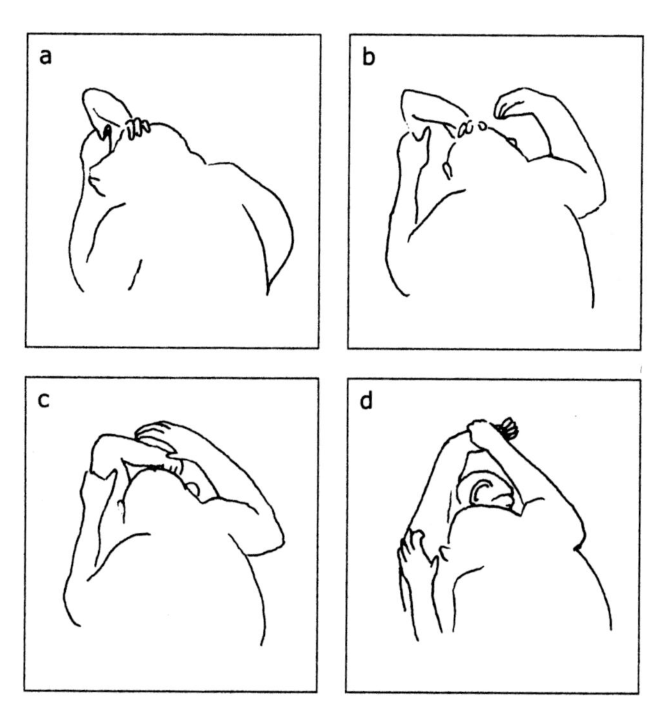

Fig. 5.3 A mother takes the initiative in holding her offspring's hand, and seems to mould hand-clasp grooming (Drawing by Michio Nakamura from video clips).

but we do not have the means to prove that chimpanzee mothers intend to mould their sons' behaviour. As evidence is lacking that mothers were deliberate in their behaviour, it perhaps should not be called moulding, but at least it is a kind of scaffolding. This is a kind of social transmission of information that is different from the usual passive observational learning.

The offspring's learning was influenced by the mother's manual actions. Since the dimensions of the torso and arms are different between mother and immature offspring, the postural configuration of GHC between mother and youngster is unbalanced and awkward. That they assume such a posture in spite of the difficult stance is surprising. The youngest chimpanzee seen to engage in GHC with the mother was six years old (males) and five years old (females).

In 2009 I found *Gwekulo* engaging in GHC – with straight elbow and palm-to-palm – with an adult female, *Xtina*, whose one-year-old infant had been cared for by *Gwekulo*. *Xtina* had not come from K-group, so it is likely that she learned the straight elbow and palm-to-palm type of GHC from *Gwekulo*, or was at least influenced by *Gwekulo*'s assumption of this style. Thus, *Gwekulo*'s influence extends beyond *Pinky*'s family and is still continuing, although it is far from being M-group's norm.

5.1.2 Social scratch

'Scratch my back, and I'll scratch yours.' Richard Dawkins, who so-titled chapter 7 of his book, *The Selfish Gene*, may be surprised to learn that chimpanzees of some groups do not scratch the backs of their companions. In the early 1990s, Bill McGrew and Linda Marchant studied the handedness of Gombe chimpanzees, recording many behavioural patterns, including social grooming. They were attentive to the details of the hand and arm movements of their subjects. When they visited Mahale again in 1996, they noticed that the chimpanzees of M-group were scratching the backs of their associates, in contrast to the lack of 'social scratch' among Gombe chimpanzees (Nakamura *et al.* 2000). I was amazed to learn that chimpanzees of some sites do not scratch the back of their companions, although they groom one another's backs and scratch themselves. 'Self-scratch' is common across all study sites, but social scratch is not. Why did this pattern emerge at Mahale, but not at Gombe? Our speculation is related to the presence of itchy plants such as buffalo bean (*Mucuna* spp.) at Mahale but not at Gombe. Chimpanzees could learn to scratch socially from the experience of being scratched by others. They could remember the pleasurable feeling of being scratched when they are itchy. They might take the groomee's perspective when they groom, given their ability to reciprocate (Nakamura *et al.* 2000).

Social scratch begins typically in older infancy or early juvenility. The youngest chimpanzee seen to do it was a two-year-old female, who, however, scratched the back of an adult male with *both hands*. The youngest male was three years old, and he also scratched the back of an adult male with *both hands*! All other chimpanzees scratched others with one hand only. So, if comparison is limited to fully fledged one-handed social scratching, then of four regularly observed female infants, one began at the age of four years, and three at age five; of four regularly observed male infants, one began at age four, one at age five, one at age six, and one did not show social scratch by the age of six years. What drew my attention was that all the new social scratchers scratched the body (usually the back) of adult or adolescent males or females, but not of their mothers. In terms of an hypothesis of perspective-taking, this is curious because infants were mostly scratched by their mothers while they were infants. If they begin to scratch others by taking another's perspective, they may well be expected to groom their mothers first. So, perhaps infants acquire social scratching by observational learning, which is one of the hypotheses proposed by Nakamura and colleagues (2000). Nakamura found that social scratch most often occurred among adult and adolescent males and between mothers and their infants. So, the presence or absence of social scratch could be related to overall social structure among local populations of chimpanzees.

5.1.3 Culture hunting

Since the 1970s, new long-term field sites of wild chimpanzees have been opened, one after another (Stumpf 2007). For example, at Bossou (Guinea) in 1976, Taï forest (Ivory Coast) in 1979, Goualougo Triangle (Republic of Congo) in 1999, and Gashaka (Nigeria) in 2002. Uganda has many sites such as Kanyawara begun in 1987, Budongo (re-)opened in 1990, Ngogo in 1994, Kalinzu in 1995, and Semliki in 1996. From these sites, more and more local traditions have emerged.

In 1997, at the invitation of Christophe Boesch, I first visited Taï forest for a short period and was astonished to see many behavioural patterns that I had never seen before. For example, travelling chimpanzees pushed down termite towers with one hand, ate soldier termites, then moved on. Some chimpanzees high in an emergent tree found a giant fruit (*Treculia*). The fruit fell to the ground, and a few chimpanzees gathered around the fractured fruit and began to share it.

Although I had never seen such behaviour at Mahale, the termite species at Taï differed from those at Mahale in terms of the shape of their termite towers, and the giant fruit eaten at Taï was unavailable at Mahale. The lack of observation was explained simply by the lack of stimulus objects. It is natural that chimpanzees respond flexibly to different environmental conditions with differing behavioural patterns.

Therefore, I decided to visit study sites where the same subspecies of chimpanzees reside and the environment is broadly similar and to document their behaviour with a video-tape recorder.

In August 2001, I visited the Ngogo area of Kibale forest, where John Mitani and David Watts had been working after John had done research at Mahale. The Ngogo research camp was very comfortable, furnished with a spacious cabin-style hut, electricity, a huge water tank storing rain water, good desks, and a refrigerator with beer. What I missed was animal protein, as all of the researchers present at that time seemed to be enjoying vegetarian meals. However, I was lucky to be able to supplement my meals with dried *Macrotermes* termites, which I had bought as a treat at a local market on the way to Ngogo.

Ngogo is proud of the largest group of chimpanzees that has ever been recorded. Its study group numbered at least 150 chimpanzees, very probably more (Mitani & Watts 1999)! John called my attention to the scarcity of pant-grunt greeting among the adult males and to the alpha male in particular. This was true: to my surprise, adult males rarely pant-grunted to the alpha male. Perhaps, the presence of so many adult males produced various male factions, which would make it possible for lower-ranking males not to be forced to express vocal gestures to the alpha male or to their factional comrades. Male chimpanzees hate to pant-grunt and are likely to skip greeting, if any excuse is available. This finding reminds us that behavioural differences may be caused by not only local traditions but also by demographic conditions.

5.1.4 Even social scratch patterns differ!

I heard that the chimpanzees of Ngogo also performed social scratch, scratching another's body with one hand. Of course, I had not expected to find any differences across sites in the behavioural patterns of social scratch. When I visited Ngogo, I immediately saw them social scratching, because it was common. However, to my surprise, it was accompanied by curious gestures and strange sounds. I watched 12 adult or adolescent males, and all of them scratched his companion's back or upper arm by 'poking' with all five fingers straightened (Fig. 5.4a, b). In contrast, chimpanzees at Mahale scratch socially by 'stroking' slowly with all fingers bent (Fig. 5.5a, b). So, the scratching motor patterns

Fig. 5.4a, b Social scratch – poke at Ngogo.

Fig. 5.5a, b Social scratch – stroke at Mahale.

were completely different, although the situation was similar (Nishida 2003b). If you consider this difference to be trivial, then you are in error. The chimpanzees of both sites engage in social scratch in the same context of social grooming, yet one of them 'strokes' and the other 'pokes'. This cannot be due to genetic differences, because at least one of the adult females of Mahale occasionally 'pokes'. The subtle difference seems to result from socially learning either style under the influence of model individuals that engage in either style.

In a brief visit to Gombe, Masaki Shimada found a few chimpanzees of Gombe who scratched socially (Shimada 2002), but the pattern was of the 'poking' variety. However, social scratch is uncommon at Gombe, as McGrew and Marchant observed long ago. So, there are groups that show the social scratch custom and those that do not. Moreover, groups that display the social scratch custom do so distinctively.

Thus, culture 'hunting' has proven to be a useful method to seek out different social customs.

5.2 COURTSHIP

Courtship is pre-copulatory behaviour in which one sex, usually the male, shows the intention to mate with the opposite sex. Adult males court an oestrous female by gazing, sitting with shoulders squared ('sitting hunch'), sitting with open thighs ('male invitation'), repeatedly stamping the ground with one foot, and shaking a branch or shrub with one hand, while showing an erect penis. Another approach is to stand bipedal and extend the arms towards the female ('bipedal swagger'). Such patterns are not only similar between Gombe and Mahale but seem to be universal across species.

However, during the course of research, we have found locality-specific courtship gestures. Courtship is typically species-specific among many animals, including birds and mammals. Therefore, when I discovered a courtship pattern that is limited to Mahale, it was a total surprise, as we tend to think it to be crucial that animals should show patterns common across a species in order to make their reproductive intentions well understood. However, from the selfish-gene perspective, an individual should take any opportunity to raise its reproductive success, so long as the intention is understood. The male chimpanzee has a long, conspicuous penis and the female a conspicuous sexual swelling, so their intentions will be obvious if visual and morphological signals are shown.

5.2.1 Leaf-clipping display

In 1971 I observed a leaf-clipping display in the chimpanzees of K-group. Typically, a male holds a leaf petiole and pulls the leafstalk between his upper and lower teeth, creating a crackly sound (Nishida 1980, 1987, 1997a). After gnashing the stem through his teeth many times, the pieces of the leaf blade fall to the ground, so the remaining midrib looks like a toothpick (Fig. 5.6a, b, c). When this happens, the male discards the 'toothpick', picks up a new leaf, and repeats the process. An oestrous female watches him and tentatively approaches him. Immediately upon noticing this approach, indicating that he has succeeded in attracting the female's attention, he leads the female away from the group and mates with her.

This is a common courtship display performed by both males and oestrous females (Nishida 1980, 1987, 1997a). Use of the same courtship pattern by both sexes is remarkable, because courtship displays

Fig. 5.6a, b, c Leaf-clip display.

(c)

Fig. 5.6a, b, c (cont.)

that are universal for a species, such as stamping or sitting hunch, are done only by males. I proposed the hypothesis that the leaf-clipping display originated as a displacement activity in a conflict situation. I imagined such a scenario: a young male is attracted to an oestrous female, while higher-ranking males are close by her. He falls into conflict between attraction to the female versus fear of the dominant males, and in spite of himself he puts a leaf between his lips and pulls it through. Luckily, she notices the ripping sound, sees his willingness to mate and approaches him.

Leaf-clipping occurs at other sites; e.g. at Ngogo both males and oestrous females showed it as courtship display, as at Mahale (Watts 2008). However, in other places, such as at Bossou and Taï, leaf-clipping is done mainly in different contexts, such as play, buttress drumming, or displacement activity (Sugiyama 1981; Boesch 1995). Tomasello and Call point out that leaf-clipping may act as a kind of 'attention-getter' (Call & Tomasello 2007). I agree with this interpretation. Leaf-clipping may have been invented as 'attention-getters' in many sites, but have become fixed as customary courtship displays as a result of ritualisation at Mahale and Ngogo.

5.2.2 Shrub-bending display

Another courtship display found at Mahale is dubbed 'shrub-bend' (Nishida 1997a; Nishida *et al.* 2009). A male pulls and bends a shrub, or pulls stems of slender bamboo grass one after another, tucking them in turn under his foot or buttocks, making a terrestrial bed-like structure out of it. This he occasionally pounds down with one hand, and then he stamps the shrub – thud, thud, thud – with one foot. Most males (but no oestrous females) have been seen to do this 'shrub-bend display' (Fig. 5.7a, b, c), but their techniques vary by the slightest bit (Nishida *et al.* 2010). For example, some males do shrub-bending in a standing (quadrupedal) posture, but other males do it while sitting.

I suspect that, like leaf-clipping, the shrub-bending display also originated as a displacement activity in conflict situations (Nishida 1997a; Nishida *et al.* 2009). Perhaps, when a female does not approach to present to him, the male becomes frustrated and is thrown into conflict. Copulation cannot proceed without the female's assent, nor can he resort to force, because higher-ranking males would rush to attack him if she screams. In spite of himself, he bends a shrub, which produces a sound that accidentally attracts the female's attention. The meaning of the signal is obvious to the female, because he simultaneously gazes at her and shows penile erection.

These two gestures produce notable sounds and are always accompanied by penile erection or maximal swelling of the sexual skin. Leaf-clipping and shrub-bending commonly occur one after the other in males. Leaf-clipping is performed by both males and females, whereas shrub-bending is done only by males. Perhaps shrub-bending would be too physically demanding a pattern for females to do

(a)

Fig. 5.7a, b, c Shrub-bend display.

(b)

(c)

Fig. 5.7a, b, c (cont.)

regularly, as unlike leaf-clipping, it is usually accompanied by ground-stamping.

Neither of these signals is shown at Gombe. If a male chimpanzee from Mahale courted a female chimpanzee from Gombe in this fashion, can you imagine what kind of response she would give? Males show penile erection and then face towards the oestrous female, so I am pretty sure the point would get across, but the female might be a trifle bewildered!

There are also completely idiosyncratic patterns. An old male named *Musa* plucked handfuls of grass or shrubbery and made a 'shooo' sound by whisking them up and down in one hand (Nishida *et al.* 2010). An adult male, *Pim*, stood bipedally, but with both hands on the ground, and walked forward towards an oestrous female, showing a fully erect penis (Fig. 5.8a, b).

Fig. 5.8a, b *Pim* stands bipedal, but with both hands on the ground, and walks forward towards an oestrous female, showing a fully erect penis.

On another occasion of 'culture hunting', Michio Nakamura visited Bossou, Guinea. There he encountered the behavioural patterns of 'heel tap', which was used in courtship, 'mutual genital touch', which was used in greetings among females, and 'index to palm', which seemed to be comparable to leaf-grooming at Mahale. Mahale chimpanzees lack these patterns (Nakamura & Nishida 2008). Since the chimpanzees of West Africa differ genetically from those of East Africa, it would also be necessary to investigate genetic differences in seeking an explanation for these behavioural differences.

5.2.3 Courtship by juvenile or infant males

A sexually immature male's courtship display is a simplified behavioural pattern in which he stands bipedal and taps on either the female's shoulder or back. If the female does not crouch on all fours, the male youngster begins to whimper. As whimpering is an infant's signal of its demands to the mother, the female immediately understands that the male in this situation is soliciting her to present for mating.

Standing bipedal and tapping the female's shoulder, jumping up and down in place, or throwing tree branches is the non-stylised, rude version of behaviour that occurs mostly in juvenile courtships (McGrew *et al.* 2001). Juvenile courtship is a mixture of 'intention movement' and 'conflict behaviour'. However, young males start to try 'shrub-bend' and 'leaf-clip' displays at about three years of age, although they sometimes do so without facing an oestrous female and even when no oestrous female is present. This kind of courtship behaviour is a type of playful practice. That a certain age and amount of experience is required for lovemaking is the same in the human world as it is in the world of the chimpanzee.

5.3 VOCAL COMMUNICATION

Frans de Waal visited Mahale in 2002 at my invitation. He is a chimpanzee expert but never before had seen chimpanzees in the wild. His strongest impression upon seeing wild chimpanzees for the first time concerned their vocal communication. He was surprised to see chimpanzees always alert to the calls of their companions nearby or distant in the natural environment (de Waal 2003). Moreover, the apes seemed to listen to all sounds in the environment surrounding them. He had not considered vocal communication to be so important, as he was accustomed to watching chimpanzees only in captivity.

Chimpanzees of Mahale staccato, whimper, scream, pant-grunt, aha-grunt, bark, hoo, pant-hoot, 'wraa', and emit other sounds. Although vocal communication is not my area of expertise, I think the calls used by Mahale chimpanzees are broadly similar to those described by Goodall for Gombe chimpanzees (Goodall 1968, 1986).

However, vocal communication also shows local differences. The first person to become aware of the differences in the sounds of pant-hoots between the adult male chimpanzees of Mahale and Gombe was Toshikazu Hasegawa, who had listened to some call samples of adult male chimpanzees at Gombe, taped by Peter Marler. John Mitani of the University of Michigan came to Mahale for the first time in 1989, and his visit presented a good chance to investigate the differences. Call samples from Gombe and Mahale were matched up using sonograms, and John attested to the differences (Mitani *et al.* 1992).

The pant-hoots of males at Mahale are high-pitched and of short duration. By comparison, pant-hoots from Gombe seem drawn-out. Incidentally, when hearing the calls of one of the males (*Musa*) at Mahale, anyone could pick out to whom it belonged. This was validated by the sonogram, which showed *Musa*'s vocalisation to be the only one that was characteristically Gombe-style.

5.4 OTHER BEHAVIOURAL DIFFERENCES

As described in Chapter 2, there seem to be differences in colobus hunting between Gombe–Mahale and Taï: more adult colobus are selected as prey targets, and meat is shared according to the proportion of an individual's contribution to hunting at Taï. If this difference cannot be explained ecologically, then it might be attributed to tradition.

Group cohesiveness is said to be high in western populations such as Bossou and Taï, while it is repeatedly stated that smaller parties are typical in eastern chimpanzees. However, it has not been shown that western groups are more cohesive. If differences exist, they may be attributable to the differences in the size of unit-groups, because smaller groups tend to remain more cohesive (Lehmann & Boesch 2004). Moreover, environmental conditions vary from site to site, and it is difficult to elucidate how these variables affect party size and cohesiveness in each study site. Traditional influences remain a possible candidate to explain these social differences until more systematic comparisons of grouping are made.

It may be that intimidation displays such as 'throw splash' (see Chapter 8) are customary at Mahale. Since juveniles show diverse displays in solicitation of play, some of which are related to threat display and aggression, we may well expect to find many traditional displays at different study sites. This line of research may yield a full range of rich intergroup differences in wild chimpanzees.

6

Female life histories

6.1 FOSTER MOTHERS AND ORPHANS

6.1.1 *Tula*: a daughter with such charm

Tula's mother is *Sada*, who was born in K-group. As described earlier, we used sugarcane provisioning to increase our contacts with the Mahale chimpanzees. My first success with individual recognition of chimpanzees through provisioning was in 1966, when *Sada* was still riding on her mother's back. So, *Tula*'s family is one of the very few chimpanzee families to be documented over four generations.

Tula is a female who is widely discussed. She has narrow eyes – not quite a beauty queen, but definitely charming. She first stole our hearts and dominated our camp conversations when she was five years old. She picked up a cigarette butt that one of my colleagues had tossed and pursed it between her lips (Please forgive us! This occurred almost 30 years ago!).

When she was seven years old, she pulled off a sign that we had secured to a *Myrianthus* tree with the message written in Swahili: 'The fruit that this *Myrianthus* tree bears is property of the chimpanzee'. *Tula* paraded around for a half day with the sign on her back. On another day, she jumped into a large puddle and began turning somersaults, laughing to herself and splashing, thrashing, and indulging in all sorts of solitary play.

By the time she was nine, she approached us and began swinging branches around; she threw sticks whenever we came close to her. The first victim she hit with a stick was Linda Turner, an American woman who had come to help with research. Another time, Miya Hamai saw *Tula* pick up a floral-patterned aluminium bowl with the bottom missing. She put it on her head, modelling it, then walked around poking her face out of the hole; finally, she placed the bowl on her belly and banged it like a drum.

When she turned ten years old in 1990, an event occurred that deeply touched all of us.

6.1.2 *Tula* adopts the orphan *Maggie*

It was the middle of the dry season on 17 August 1989. Kenji Kawanaka and a research assistant, Moshi Bunengwa, noticed a young adult female, *Pinky*, carrying an infant on her back, although it was known that she did not have any children at that time. They thought this to be bizarre, so they moved in for a closer look and found the infant to be *Maggie*, the baby of *Wantendele*. *Maggie* had injuries to both her left foot and left hand. *Pinky* dumped *Maggie* when three adult males made a charging display at her. Just then, *Maggie*'s big brother, *Masudi*, came plodding up the path and, seeing *Maggie*, put her on his back, protecting her from the males' display.

We confirmed that *Maggie* and her mother had been sighted together on 11 August 1989. So what became of *Wantendele*? From 14 August, a lion had been prowling around the research camp, so the possibility was great that the lion had taken her as prey (see Chapter 2). *Maggie* was next seen on 6 September, and during those three weeks no one had a clue who had been caring for her. *Maggie* was exactly three years old. She had not been weaned yet, and at that age, a mother's breast and protection were considered necessities.

When she re-appeared on 6 September, *Maggie* was riding on ten-year-old *Tula*'s back. Despite *Tula* being only ten, her level of care was far better than that shown by *Masudi* or *Pinky*. She had adopted *Maggie* to be her child. If *Maggie* strayed even a few metres out of *Tula*'s sight, other juveniles hastened to bully her. Every time, *Tula* came flying to *Maggie*'s rescue.

I was in Japan when Moshi wrote to me: '*Tula* and *Maggie* are like mother and daughter. *Tula* has an extremely good character, like that of a human mother.' I was anxious to see *Tula* and *Maggie*.

The privilege of seeing the pair came about a year after *Maggie* was orphaned. They certainly looked like mother and daughter, with piggy-back rides, grooming sessions, hours of play, and food sharing – and if any other juvenile picked on her, *Tula* would hasten to the spot to hold and console little *Maggie*. At night, *Tula* would make a bed far greater than anyone else's and the two of them would fall asleep. *Tula* fulfilled all the roles of a natural mother, aside from breastfeeding *Maggie* (Fig. 6.1).

However, in comparison to her peers, *Maggie*'s coat was wiry. It seemed that *Tula*'s grooming left a little to be desired. Furthermore,

Fig. 6.1 *Tula* invites *Maggie* to ride dorsal for departure.

before *Maggie* was orphaned, her physical size had been above average, yet a year after her adoption, her small size stood out; the lack of breastfeeding had taken its toll.

Tula was now 11, and old enough to be in oestrus. Once in oestrus, *Tula* began to give *Maggie* less and less attention, neglected her frequently, and ranged without her, just as a real mother would. *Maggie* began to whimper regularly. Despite not nursing, it seemed that *Tula* was making an effort to 'wean' *Maggie*. It became a common thing for *Tula* and *Maggie* to split up during the afternoon. *Maggie* went foraging on her own for fruit, falling behind her companions, and at times letting out a little scream and running to catch up. But every evening *Tula* and *Maggie* met up to share the same bed.

About 18 months after *Tula* had adopted *Maggie*, both of them disappeared. A month later, *Tula* and *Maggie* just turned up, together. It was obvious that *Tula* had taken *Maggie* to visit a remote part of the range, perhaps one of the neighbouring groups.

Two years had passed since *Tula* adopted *Maggie* when I was again at Mahale; I noted that *Tula* had ceased her care of *Maggie*. *Maggie* was five, and even if her natural mother were still alive, it would have been *Maggie*'s weaning period. Until *Maggie* lost her mother, her growth and development had been rapid. Even with having breast milk cut off at the age of three, she had undoubtedly acquired the will to live. But it is

Fig. 6.2 *Maggie* and *Tula* rest together.

highly questionable as to whether or not she would have survived had *Tula* not become her guardian (Fig. 6.2).

Alloparental behaviour is most commonly seen in females, starting in juvenility and continuing through adolescence, and is said to be highly effective as practice for later maternal behaviour (Nishida 1983b). Even so, *Maggie* was extremely lucky to have the good fortune of a devoted foster-mother.

6.1.3 *Gwekulo*'s foster children

Gwekulo immigrated into K-group as an adolescent female in 1972. Her first opportunity to care for an infant of another female came in 1976. A primiparous mother, *Chausiku*, caught a cold and was so emaciated that she abandoned her obligation to care for her own infant, *Katavi*. *Gwekulo* adopted *Katavi* for a week or so, until *Chausiku* recovered from the illness (Uehara & Nyundo 1983).

When K-group disintegrated (see Section 6.2), *Gwekulo* immigrated into M-group, along with many other K-group females. *Gwekulo*'s intimate relations with *Chausiku* continued, and the latter's second son, *Chopin*, was cared for by *Gwekulo* in 1987. Meanwhile, *Gwekulo* began to associate closely with *Wakasunga*, another female immigrant from K-group, and to care for her juvenile son, *Linta* (Fig. 6.3). When *Linta* grew up to be a

Fig. 6.3 *Gwekulo*, *Wakasunga*, and *Linta*.

Fig. 6.4 An orphan, *Pipi*, eats fruits of *Marantochloa*, which most chimpanzees ignore.

juvenile around 1989, *Gwekulo* was seen to interact with many infants. In 1993 *Gwekulo* often accompanied the alpha female, *Wakampo*, and cared for her one-year-old son, *Brutus*. In September–October 1993, the first flu-like disease hit M-group (Hosaka 1995a) and killed a primiparous mother, *Pulin*. In 1994 *Gwekulo* adopted the 45-month-old orphan, *Pipi* (Fig. 6.4); she provided all the maternal care except for lactation to *Pipi*.

Fig. 6.5 *Gwekulo* and *Pipi*.

So, to casual observers, they appeared to be a genuine mother–infant pair (Fig. 6.5). (By that time, we had realised that *Gwekulo* was sterile.) Throughout the juvenile period, *Pipi* received grooming and social support from *Gwekulo*. Thanks to *Gwekulo*'s dedication, *Pipi* grew up to be a very active juvenile female who engaged in social play with males of comparable age. After *Pipi* reached adolescence, she and *Gwekulo* tended to travel separately, but occasionally when *Pipi* presented her back to *Gwekulo*, the older female immediately groomed her. However, in September 1995, *Gwekulo* became a babysitter for *Pinky*, who had immigrated to M-group in 1983. *Gwekulo* was often observed carrying *Pinky*'s four-year-old infant son, *Primus*. *Gwekulo* and *Pinky* became a formidable coalition. No other adult female dared to threaten either female or *Primus*, because in almost all cases she would have to contend with this strong coalition – any group member in conflict with either of them had to fight both of them. When *Pinky* gave birth to another infant, *Puffy*, *Gwekulo* shifted her attention to *Puffy*, perhaps because *Primus* became less attractive.

Pinky died of another flu-like epidemic in 2006 (Hanamura *et al.* 2008), since which time *Puffy* has spent most of her time with *Gwekulo* as a substitute for her mother. As a childless female, *Gwekulo* has dedicated her whole life to raising infants of other females. I think that the maternal instinct is so strong that a sterile female cannot help but direct her passion to biologically 'incorrect' targets. Although there is little biological cost to alloparenting for a sterile female, her dedication strikes me as a grand achievement.

6.2 THE EXTINCTION OF K-GROUP

We must go back to an earlier stage of our study (1979–1985). In our research of Mahale chimpanzees, we cannot leave this topic untouched.

6.2.1 Strained relations between two groups

Although Goodall began her research five years earlier than our Mahale project, how did I infer first that chimpanzees form a large social unit (called the 'unit-group' or 'community') and that relations between social units are antagonistic, and that females transfer between unit-groups?

From the outset we felt confident that the chimpanzee's social unit ought to be more complex than just mother–offspring relationships. Our belief that chimpanzees have larger social units than a mother–offspring pair also resulted from previous field studies at Kabogo, Kasakati, and Filabanga (Azuma & Toyoshima 1962; Itani & Suzuki 1967; Suzuki 1969; Izawa 1970; Kano 1972). We observed what we called a 'large-sized group' in these sites. I clearly recollect seeing 32 chimpanzees travelling in a straight line, moving from one riverine forest block to another, in the savanna terrain of Kasakati in August 1965, before I began my research programme at Mahale.

My discovery of the social unit was partly due to coincidence. I established the sugarcane plantation and feeding site in the overlapping area of two groups, without any precise knowledge of chimpanzee social structure. If I had established the feeding site at the centre of M-group's territory, for example, I might have spent many more years without recognising the social unit and the significance of intergroup hostility.

It is now well-established that chimpanzee unit-groups are mutually antagonistic. However, they do not fight at random. Adult males and a few females occasionally patrol the periphery of the group range (Nishida 1979). After letting out a hearty pant-hoot chorus at the boundary of the range, they keep silent for a bit, awaiting a response from the neighbouring group. If an outburst of pant-hoots is returned, they rush back towards the centre of their home range.

As shown in Fig. 6.6, K-group's range greatly overlapped M-group's range. In the 1960s and most of the 1970s, mostly around September,

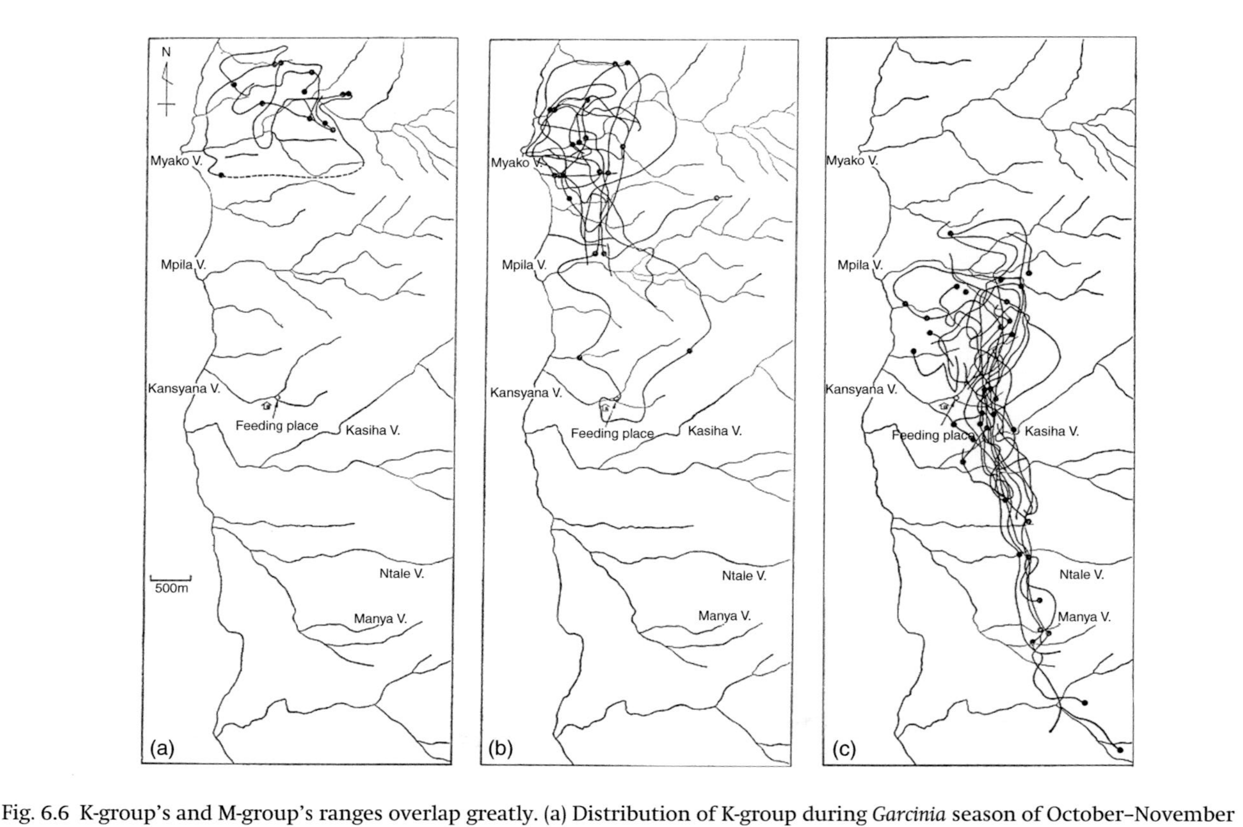

Fig. 6.6 K-group's and M-group's ranges overlap greatly. (a) Distribution of K-group during *Garcinia* season of October–November 1967. (b) Distribution of K-group during *Saba* season of November–December 1971. (c) Distribution of M-group during *Saba* season of November–December 1966. ● = sleeping site (from Nishida 1974).

Fig. 6.7 Myako camp, when it was thriving in 1979.

M-group formed large parties, headed northward, and used the overlapping area, whereupon K-group immediately moved northward to the Myako valley (Fig. 6.7), evacuating to a 'safe' area. K-group remained around the Myako valley until the following February or so, before returning to the overlapping area that included the Kansyana provisioning site.

Depending on the year, there were occasions when K-group did not evacuate to Myako valley. During these periods, both groups were sometimes within 500 m of each other, but as K-group avoided encounters at all costs, usually nothing happened.

Immediately after K-group left the provisioning site, M-group showed up. M-group chimpanzees sniffed the ground at the provisioning site and discharged loose stools. M-group chimpanzees then headed north and, as they closed in on K-group, M-group members also voided diarrhoea. Most likely the scents of unfamiliar chimpanzees stirred up great emotional tension in them, bringing on diarrhoea.

An aged K-group female, *Wamikambi*, and her infant son, *Limongo*, once failed to escape from the provisioning site and were beaten to within an inch of their lives. *Wamikambi* clasped her son under her belly and he was able to cling to his mother throughout the assault. Researcher Mariko Hasegawa later said that our research assistant, Ramadhani Nyundo, and his crew waved sugarcane around the attackers and rescued the mother and infant (Nishida & Hiraiwa-Hasegawa 1985).

6.2.2 *Wantendele*

In another case, we observed an oestrous K-group female, *Wantendele*, attempt to immigrate into M-group with her son, *Masudi*, who was in mid-weaning. But her oestrus was not full-fledged. A group of adult males launched an attack on her, one after the other. A huge crowd, including M-group's adult female chimpanzees, gathered and let out shrieks, adding to the chaotic scene.

I had a feeling that the mother and child were about to be slain, so I hollered to the students for help. We enclosed the mother and child, and thwarted the attack. Although *Wantendele* and her son were trapped in the middle of a human circle, they remained incredibly calm, not falling into a terror-stricken panic, nor even expressing fear in the slightest. Perhaps she understood our good intentions, even if we were a different species! Most amazing was that once we prevented the attack the chimpanzees' frenzy ceased immediately, and after that *Wantendele* and her child were not harassed (Nishida & Hiraiwa-Hasegawa 1985).

Some might claim that our intervention goes against the basic 'rules of science' that dictate that research ought to be conducted without any human intervention into an all-natural habitat. I forgot this rule. I rescued the mother and child before I remembered that I should not intervene, because *Wantendele* was like one of my long-term, human friends. Even if I had remembered the 'rules', it would have been more important for me to keep *Wantendele* and her child alive than to obtain solid evidence of murder by adult chimpanzees.

After our paper had been published reporting this incident, Birute Galdikas, a famous researcher who studies orangutans, sent me a short telegram-like letter. In it she wrote out, all in capital letters: 'YOUR TEAM WAS CORRECT IN RESCUING MOTHER AND CHILD. I'M BEHIND YOU 100 PERCENT.' At first I had no idea what this message referred to, and then it struck me that she was talking about the incident with *Wantendele* and her son. Perhaps some Western primatologists claimed that Japanese researchers were interfering unnecessarily with wild chimpanzee behaviour. Galdikas may have thought that we had received such criticisms.

Wantendele and her son retreated to the safety of Myako valley, but after some time she again tried to join M-group. She was attacked again, but it was not a joint attack, and even when one male attacked her, other males came to her rescue.

6.2.3 K-group becomes extinct

When first counted, there were six adult males in K-group, but during a six-year interval from 1969 onwards, four adult males vanished, one at a time. The group lost the oldest adult male, *Kasagula*, in 1969; in 1970 a young prime adult male, *Kaguba*, the lowest-ranking male, disappeared for unknown reasons. In April and October 1975, two males, one prime (*Kasanga*) and one past-prime (*Kajabala*) disappeared from K-group, again for unknown reasons. Since the adolescent male, *Sobongo*, had grown into adulthood in 1971, there were three adult males in K-group in early 1976: *Kasonta*, the alpha male, *Sobongo*, the beta male, and *Kamemanfu*, the gamma male.

Thus, K-group comprised three adult males, nine adult females, and ten immature individuals. Within the next five years, two adult males, *Kasonta* and *Sobongo*, disappeared, with the last remaining adult male, *Kamemanfu*, vanishing in 1983.

We deduced that at the very least, M-group had killed most of K-group's adult males. The reasoning behind this was that the males who vanished were in good health whenever we observed them, but that every time K-group and M-group had an encounter, adult males from K-group vanished. The day before *Kasonta* vanished, Shigeo Uehara confirmed that a ruckus was heard coming from the foothills of upper Myako, although the steep terrain prohibited him from approaching the chimpanzees.

From 1977, when there were only two adult males (*Sobongo* and *Kamemanfu*) left, K-group's adult females gradually began to migrate into M-group. In 1979, when *Kamemanfu* was the last adult male remaining, the rest of the female cluster made a mass migration. By the end of December 1979, all the cycling adult females, parous or nulliparous, with children at post-weaning, made a beeline for M-group's range. The non-cycling females left behind followed suit after their young were weaned, from 1980 to 1983. This is how the extinction of K-group became an established fact in Mahale's history.

This saga tells us that male chimpanzees forming the core of the unit-group compete for territory, and that females select a dominant unit-group that has many adult males. Females are not passive subjects in the society. No females were abducted by the adult males of M-group – rather, the females emigrated from K-group without coercion (Nishida *et al.* 1985).

After the extinction of K-group, M-group extended their home range to the north, and B-group, who were further north, extended their home range to the south, with both groups settling on Myako valley as the border. No one who visits Mahale today can imagine that at one point, in their heyday, K-group had 30 members whose unit-group ranged between B- and M-groups.

Although no lethal attack of K-group by M-group was witnessed, such attacks accompanied by expansion of the territory of an aggressive group have recently been seen at Ngogo, in the Kibale forest of Uganda (Mitani *et al.* 2010).

6.3 IMMIGRANT AND RESIDENT FEMALE RELATIONSHIPS

6.3.1 Newly immigrated females get picked on

Most females emigrate from their home group at around the age of 11 years. We have documented three types of female: one who emigrates and never returns to her natal range; one who emigrates but then returns once or several times but finally emigrates; and finally one who remains in her natal group without ever leaving, or who may leave for a brief period but then returns and stays. The second type of female is most common, while the third type is rare. Whether it is temporary or permanent, when a female emigrates from her natal range she is in oestrus. In general, parous females do not transfer to other groups.

Resident females pick on immigrant females (Nishida 1989; Kahlenberg *et al.* 2008; Pusey *et al.* 2008). I have dubbed this scenario the 'evil mothers-in-law'. This is because it was said that in traditional Japan, a newly wedded wife was often bullied by her mother-in-law. Here, the 'daughters-in-law' (the immigrant females) seek protection from adult males and almost never leave their side. When the adult males see the resident females attacking immigrant females, the males tend to take the side of the immigrant females or interfere and stop the assault. At times, the adult males assist the resident females, but as there are always males there to protect the immigrant females, no immigrant females have been lost in these interactions. Whenever one of the immigrant females is alone she regularly lets out a loud pant-hoot, indicating her location to one of the adult males who may act as a bodyguard (Kahlenberg *et al.* 2008; Pusey *et al.* 2008).

Close relationships sometimes form among resident and immigrant females, similar to what is reported for bonobos (Idani 1991). An immigrant female targets a high-ranking resident female and begins to follow her incessantly. The immigrant female grooms the high-ranking resident and tries to curry favour with her. Eventually, the resident female allows the young immigrant to babysit her infant, and the immigrant female spends most of her time with this particular mother and infant. This relationship offers three advantages to the immigrant: first, she is protected from harassment by other resident females and from predators; second, she gains knowledge of food distribution; and third, she practises maternal care. For resident females, advantages include being groomed, getting a babysitter, and protection for her and her offspring. The cost to her of decreasing food resources may not be great, because she is high-ranking and may occupy a core area of high food productivity (Idani 1991).

The root cause of the aggression to newly immigrated females is that most resident females are unwilling to share their food resources with females coming from a foreign range. However, in reality, the resident females rarely pull together to stage an attack on immigrant females. Why is this so?

Over the course of a year, an adult female uses the entirety of the group's range, but she also claims a certain area as her 'turf'. This space is called the 'core area'. Each M-group resident female's core area is in a slightly different location; this spacing is not really a question of cooperation, but rather the opportunity to compete is reduced to begin with.

In addition, consider this situation from a resident female's point of view: if an area that is often exploited by an immigrant female is part of that resident female's core area, the cost to her is heavy, but if part of some other resident female's core area is targeted, then it is no big deal. Since every resident female tends to overlook an immigrant female using someone else's core area, in the end, a cooperative attack by resident females against a newcomer is rare.

For adult males, on the other hand, rather than a decrease in share of food, they are more concerned with the potential for an increase in mating partners (Trivers 1972). Moreover, adult males, unlike females, do not stake such claims on core areas, suggesting that there is far less direct foraging competition for them to contend with. Given these conditions, most adult males warmly welcome immigrant females.

Because it is such an ordeal for an immigrant female to establish her own core area, several years pass before she becomes pregnant and gives birth. It took *Nkombo* (an original immigrant female from K-group) eight years to finally establish a core area in the southern sector of M-group's range. It was 12 years after transferring that she had a child. Hers is an exceptionally lengthy case, but according to 30 years of data, the median number of months before giving birth after transferring is 32 (Nishida *et al.* 2003). Moreover, the first child usually dies. It seems that younger females are susceptible to miscarriage. Furthermore, if a newly immigrated female gives birth, her newborn infant is at high risk of being the victim of infanticide (see Section 6.4).

6.3.2 Why do females transfer?

As seen in Chapter 3, mother–adult daughter relationships are always warm, and these females may form strong coalitions. So, continued natal group residence confers a great advantage on a female who does not emigrate. In spite of this, more than 90 per cent of females emigrate from the natal group (Nishida *et al.* 2003) and immigrate into a group where no maternal support is available. Why?

Such emigration may occur because the cost of natal residence usually exceeds its benefit. The costs of natal residence are inbreeding and indirect feeding competition among close relatives. Although we have never seen copulation between brothers and sisters in 40 years of research (Nishida 2001) (Table 6.1), father–daughter incest may occur if females do not emigrate. Such inbreeding would dramatically decrease a female's reproductive output. At Gombe, mother–son mating occurs and an offspring was born (Goodall 1986; Constable *et al.* 2001), but at Mahale we have had only one case of mating by an adult male with his mother and one case of attempted copulation by a juvenile male with his mother. In the first case, the copulation was completed but no offspring resulted. In the latter incident, the mother resisted and the mating was unsuccessful (Nishida 1990b). Thus, incest among close relatives is virtually absent at Mahale.

Due to epidemics that struck M-group several times (Chapter 10), it greatly decreased in size from the maximum of about 100 to about

Table 6.1a *Copulation between older brother and younger sister in Mahale's M-group (from Nishida 2001)*

Older brother (*Br*)	Younger sister (*Si*)	Date *Si* sexually matured to approximate date of emigration/death	Approximate date of *Br*'s death	Period (years) in which sexually mature siblings co-resided	*Frequency of copulation between siblings during *Si*'s adolescence	Frequency of copulation between siblings during *Si*'s adulthood	Notes
Bembe	*Bunde*	1987–November 1991	December 1995	4	0	0	
Masudi	*Maggie*	July 1997–March 1998	June 2006	0.8	0	*Si* emigrated	
Toshibo	*Abi*	January 1992–	December 1995	4	0	*Br* died	
Shike	*Wasiwasi*	1988–April 1988	January 1992	0	0	*Si* emigrated	
Nsaba	*Tula*	May 1990–September 1994	January 1997	4.3	0	0	
Alofu	*Ai*	September 1998–June 2002	Alive	3.8	0	N/A	For two years *Ai* was sick

Table 6.1b *Copulation between older sister and younger brother in Mahale's M-group (from Nishida 2001)*

Older sister (*Si*)	Younger brother (*Br*)	Date *Si* sexually matured to approximate date of emigration/death	Approximate date of *Br*'s death	Period (months) in which sexually mature siblings co-resided	Frequency of copulation between siblings during *Br* juvenility	Frequency of copulation between siblings during *Br*'s adolescence	Frequency of copulation between siblings during *Br*'s adulthood
Penelope	*Pim*	September 1989–November 1989	February 1997	0	*Si* emigrated	N/A	N/A
Wantunpa	*Bonobo*	May 1985–December 1987	November 1990	0	*Si* emigrated	N/A	N/A
Katyentye	*Bembe*	1977–November 1995	1991	4	0	0	0
Kantamba	*Alofu*	1977–October 1981	1991	0	*Si* emigrated	N/A	N/A
Ruby	*Orion*	May 1996–July 2006	June 2000	6	0	N/A	N/A

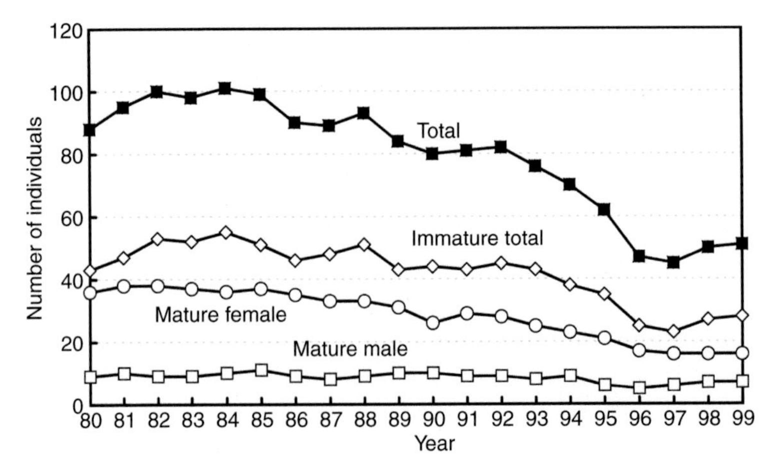

Fig. 6.8 Changes in number of mature males, mature females, immature individuals, and size of M-group (from Nishida *et al.* 2003).

50 individuals between the late 1980s and the mid-1990s (Fig. 6.8). This decline in group size seemed to influence the movement pattern of natal females: more natal females stayed with M-group. Of those females who were born between 1981 and 1988, three (*Maggie*, *Sylvie*, and *Cocos*) emigrated, but as many as five females (*Ako*, *Abi*, *Totzy*, *Ruby*, and *Ai*) did not. In 30 years of M-group demography, we have only six cases of females who remained in the unit-group after reaching full maturity, and five of the six cases were females born between 1981 and 1988.

First, with regard to incest avoidance, as many as six adult males and two adolescent males disappeared between 1995 and 1997. These absences may have decreased the possibility of father–daughter incest and sister–half-brother incest.

Second, lower population density probably weakened intra-group feeding competition and increased the benefits of remaining in the natal group. Lower risk of incest and higher nutritional benefit may have made it possible for natal females to remain in the group during this particular period.

I have long wondered why many females at Gombe do not transfer, whereas most females do so at Mahale (and also at Taï [Boesch & Boesch-Achermann 2000] and Ngogo [Mitani *et al.* 2002]). This difference may be explained by the low number of groups (three) at Gombe and by the ranges of the Gombe groups being too narrowly limited by the artificial boundary of the national park, meaning females cannot find appropriate groups for transfer.

6.4 INFANTICIDE AND CANNIBALISM

6.4.1 Chimpanzee infanticide followed by cannibalism

When Yukimaru Sugiyama discovered infanticide among hanuman langurs (Sugiyama 1965), primatological circles regarded it as an aberrant behaviour and therefore unimportant. Far from being abnormal, infanticide proved to be a regularly observed phenomenon amongst langurs, in which an outside male took over the leadership of a one-male–multi-female troop. New alpha males kill infants younger than six months of age in order to make the mothers resume oestrus and ovulate. This has proven to be an evolutionarily stable strategy (Hrdy 1979). Infanticide is now regarded as one of the most important factors influencing primate social relationships (van Schaik 2000).

Cases of cannibalism, rather than infanticide, of chimpanzees were first reported from Budongo and Gombe (Suzuki 1971; Bygott 1972). Infanticide accompanying cannibalism was regarded as predatory behaviour directed to conspecifics (Hiraiwa-Hasegawa 1994). This line of reasoning was based partly on Goodall's observations of a high-ranking female chimpanzee killing and eating babies of low-ranking females (Goodall 1977).

However, in some cases infants are killed but not eaten (Hamai *et al.* 1992). In these cases, what is the reason for infanticide?

6.4.2 Infanticide also occurred at Mahale

At Mahale, there have been at least eight cases of infanticide or cannibalism to date (Hamai *et al.* 1992). There have been only three cases in which the killer was confirmed, and always an adult male was the culprit. The other five cases were observed only as the infant was being eaten.

Yukio Takahata was the first person to witness a killing. In July 1983, a few adult males, including the alpha male, ganged up on *Wantendele*, who was trying to immigrate into M-group from K-group. She was carrying her newborn on her belly, accompanied by her weaned son, *Masudi*. The gang of males snatched the baby from her, then killed and ate it (Takahata 1985).

The case that I witnessed was, unfortunately, one I thought might occur given the circumstances of the female in question. Because *Chausiku* was one of the original immigrants from K-group who gave birth in M-group's range, I predicted that her baby might be killed. I followed *Chausiku* as my target to investigate whether or not my prediction was correct. On the morning of 15 December 1983, when Kenji Kawanaka and I started our observations, a young adult male had just begun to eat a baby. We did not see the killing, but as the only other individual in the area besides *Chausiku* and her baby was a low-ranking female, it was unmistakable that the male eating the baby was the killer (Nishida & Kawanaka 1985).

On 3 October 1989, Tamotu Asou and Miho Nakamura from ANC Productions captured an infanticide and cannibalism episode on videotape. On that ghastly video recording, the second-ranking male at the time, *Kalunde*, first snatched the baby of *Milinda*, a young immigrant female. The baby was then passed to the fourth-ranking male, and finally it reached the hands of the alpha male, *Ntologi*. He mangled its face and killed the infant. After that, the usual meat-eating carried on, with *Ntologi* in the centre, divvying up scraps. Close to 20 members of M-group feasted on the baby's flesh and bones (Hamai *et al.* 1992).

These episodes of infanticide share a number of features. They took place within five years of the victim's mother immigrating into M-group. Also, the mother's core area was on the periphery of the group's range. The baby was usually the first-born of a young, immigrant mother, and always less than one year old. After the killing, except for a few cases, the baby was eaten. In no case did the victim's mother join in the cannibalism.

Most peculiar is that it is highly likely that the infants' fathers were M-group males. There was no evidence suggesting that the mothers temporarily migrated to another group and copulated there.

6.4.3 Attacking a wanton female?

On 21 October 1985, the alpha male, *Ntologi*, and his ally, *Lubulungu*, attacked a cycling female, *Gwamwami*, in the northern part of M-group's range. She was an immigrant from K-group and tended to range in the old overlapping area of both groups, even after her immigration. *Gwamwami* had not been seen for many weeks, so we imagined that she had been travelling through the original K-group range. Why did the males attack her? I speculate that the males were infuriated by *Gwamwami*'s absence, and most especially worrisome for the males was that she was found near the boundary of the territory after her long absence. Thus, she may have had many opportunities to mate with non-group males. Perhaps this assault resulted from the same root cause of killing newborn infants born to mothers who used to travel at the periphery of the group's territory.

6.4.4 Why kill the progeny?

The pertinent question here is: why would the M-group males kill the baby of a female with whom they had mated? In other words, why do they kill their progeny? At one time I thought it might be that an adult female was the killer, and that the male had just poached her prey. If that were the case, it would be more understandable, because of the lack of blood relations between the killer and victim. But sometimes, fact is stranger than fiction.

Even in cases in which a killing is not seen, if we piece together circumstantial evidence, it is fairly certain that adult males were the killers. For the moment, the best hypothesis we have is as follows:

It is always immigrant females who are attacked, and they have no alternative but to make their core area on the periphery of the group's range because of harassment by resident females. So the place where adult males often meet these females is on the periphery of the group's range. As a result, the adult males mistake the immigrant female's baby for the offspring of a rival group's male, and kill it.

I am not completely satisfied with this hypothesis. I suspect that there are far more cases of infanticide occurring than we have been able to observe. Indeed, there are several cases when a mother is carrying a perfectly healthy baby one day, and then a few days later she appears without the baby. Is it my misconception that there are this many babies being killed?

It is not impossible that one male from M-group kills another M-group male's baby, but if this were the case, it would be every baby's destiny to die young, which is outrageous. However, a related hypothesis might be that the baby of a male who is not well-integrated with the group's adult male cluster (e.g. an adolescent male) gets killed by members of the central cluster (Hamai *et al.* 1992). This hypothesis cannot yet be rejected. I suggest this alternative because infanticide and cannibalism were rarely seen after the ex-alpha male, *Ntologi*, died in 1995. Simultaneous with the disappearance of the strong male cluster, infanticide and cannibalism in M-group seemed virtually to disappear.

Alternative hypotheses such as, 'weaklings get killed' or 'killing merely for the purpose of meat' and so on have also been proposed (Hiraiwa-Hasegawa 1994), but as it now stands, I am most convinced by my initial hypothesis.

Whether or not the hypothesis is supported, it is still important to know if the baby who was killed was the son of an M-group male. Hiroyuki Takasaki has extracted DNA samples from hair strands and food wadges of many M-group chimpanzees, but unfortunately it was not possible to collect a sample from a baby less than one year old.

6.4.5 Immigrant females increase in rank

An immigrant female's status begins to increase as the years pass, because resident, higher-ranking, elder females of the group eventually die. By then the immigrant has more friends among the resident females, and she can establish for herself a core area that is much more central in the group's territory. The threat of infanticide fades, and child rearing goes more smoothly for her.

6.4.6 Dead-baby snatcher

I must note an extraordinary event in which an adult male snatched and ate the fresh carcass of a dead baby chimpanzee. This account highlights the complex features of chimpanzee infanticide.

One morning in 1992, *Garbo*, the 41-month-old daughter of *Gwamwami*, died in front of us. The cause of her death was aggravation of an injury accompanied by the loss of her foot. Several chimpanzees took a great interest in the tiny corpse and came closer to look at it. There were five adult males, one adolescent male, and two juvenile males. The alpha male, *Ntologi*, groomed *Gwamwami* for more than two minutes. Then, adult males displayed and the chimpanzees were scattered. *Gwamwami* avoided this upheaval, cutting herself off from the party and taking shelter in a thick grove of trees with the baby chimpanzee's body. The fifth-ranking adult male, *Nsaba*, followed quietly after her, and sat 5 m away from her. After about 10–15 minutes, once *Gwamwami* had calmed down, he moved closer and began to groom her back.

I was touched by *Nsaba*'s tender-heartedness. It appeared that he was consoling the bereaved *Gwamwami*. Seven minutes later, however, *Nsaba* suddenly stretched out his arm, pushing *Gwamwami* in the back. When the offspring's dead body fell to the ground, he snatched it and darted off, leaving us at a loss for words.

We could do no more on that day, because *Nsaba* ran away so frantically. The following morning, I asked my assistants to look for *Nsaba* and to collect his stool by any means, because I realised that he was interested not in consoling *Gwamwami*, but rather in the meat of her deceased infant. Finally, we found chimpanzee hair in one of *Nsaba*'s faecal samples. *Nsaba* must have pondered his idea of stealing and eating the body long before his sudden assault. He was waiting until he began to groom her, and concealed his intentions during seven minutes of grooming. He must have been waiting for the opportunity to take the body, waiting until the mother became so relaxed that her caution disappeared (Nishida 1998).

The alternative parsimonious hypothesis is that as he groomed her, his hunger for meat increased gradually until he at last made his grab. This possibility is unlikely. First, *Nsaba* was grooming *Gwamwami*'s back from the onset of the incident, and from that position, he could not see the body that was in the lap of the mother. Second, the infant's body would give off no odour within one hour of death. Third, if *Nsaba* had no initial intention of eating the body, he must have groomed *Gwamwami* in order to console her. If so, we must assume a sudden change of mind to move from affection to attack. This only increases the puzzle. Fourth, *Gwamwami* did not take any precautions against the snatching. Before she entered the bush, she was first groomed by the alpha male, *Ntologi*. At that time, she occasionally put her charge beside her on the ground. It would have been far easier to grab it then, rather than if it had been in the mother's lap. Therefore, it is likely that most males did not consider the body as meat. Dead bodies of conspecifics are rarely eaten. This rare set of events shows how deception occurs (Trivers 1972).

6.5 FEMALE FRIENDSHIPS

6.5.1 Male and female socio-spheres

Male chimpanzees seem to be more similar to human males than to conspecific female chimpanzees, in the same way that female chimpanzees are more similar to human females than to conspecific male chimpanzees. Of course, not in facial features, but rather in behavioural patterns. It is as if men and women are distinct species! Or, at least this was a point I jokingly explained in a congratulatory speech I once delivered at a junior colleague's wedding reception – we should not think of mutually understanding one another as a riddle.

A female chimpanzee's social sphere is narrow. A mother chimpanzee with dependent offspring often spends her time alone. If she has two or three children, most social interaction is confined to her family. She may have one or two close female friends. She has long-lasting relationships with these few close friends.

On the other hand, males interact with all kinds of other males. Male interaction is both opportunistic and tactical. Out of self-interest, there are tendencies toward changing allies at will, showing a 'yesterday's friend, today's foe' kind of mentality (de Waal 1982; Nishida 1983a). Rivals, for instance, groom one another with great zeal, aside from those times of fierce competition. In most cases, they quickly patch things up, even if they have a scuffle. Males also mingle with all types of females, and groom any female they meet, at least briefly.

Resident females – that is, elder females who have stayed in the group for a long time – occasionally unite to attack immigrant females (Kahlenberg *et al.* 2008; Pusey *et al.* 2008). The only other time females unite is when adolescent or young adult males launch attacks on them for no obvious reason. Males at this stage of life are usually just testing their power over females and enjoy watching the females squirm. If push comes to shove, the females launch a counter-attack on the males while letting out a frightful chorus of shrieks. Then the males vacate the scene and retreat. However, if the males are serious in their attack, the females scatter.

6.5.2 Social relationships between females

I do not know how females become intimate friends. But there are at least four combinations of females who usually stick together. The first is females from the same group. The second, as already introduced in Chapter 3, is a liaison between a newly immigrated female and a prime, higher-ranking resident female. A young mother with an infant and the newly immigrated female can also be congenial, so long as the young mother depends on the latter for babysitting. Two newly immigrated females also may become friends (Nishida, unpublished observations). The last combination is a mother and her adult daughter who has not emigrated.

Six females who transferred from K-group to M-group at around the same time in 1979 were relatively friendly. Three of these females stuck together and were observed in each other's company on a regular basis; all of the members originating from K-group always kept a little distance between themselves and other members of M-group. In addition, the members from K-group had an alliance, and repeatedly drove other females from their fruit tree.

Chausiku, a female from K-group whom I tracked regularly, spent most of her time with her senior, *Wakasunga*. The two of them had children who were almost the same age. *Chausiku* groomed *Wakasunga* frequently, and when the pair was ranging freely, *Chausiku* followed *Wakasunga* like a dog follows its master.

Chausiku had another friend named *Gwekulo*, who was barren (see above). In this relationship, *Gwekulo* took the role of *Chausiku*'s lackey. Their relationship was one-way, and *Gwekulo* always groomed *Chausiku*; it was rarely the other way around, and if it did occur, it was brief. But there was a greater underlying reason that *Gwekulo* stuck so close to *Chausiku*. The childless *Gwekulo* took an extreme liking to *Chausiku*'s son, *Chopin*. She always gave him rides on her back, shared food with him, and when other youngsters teased *Chopin*, she was his trusted reinforcement. We called *Gwekulo* 'wet nurse', even though she did not lactate.

We have all experienced the 'two's company but three's a crowd' phenomenon before. After playing happily together as a trio of friends, one is suddenly left out in the cold, while the other two are elsewhere, being good friends. *Wakasunga*, *Chausiku*, and *Gwekulo* displayed this relationship. When the three of them were together, *Wakasunga* and *Chausiku* indulged in grooming, while *Gwekulo* was given the cold shoulder.

Even after *Chausiku* died, and *Chopin* was orphaned, *Gwekulo* still came to his rescue if the older juveniles bullied him (see Section 6.1.3 above). Gradually, *Gwekulo* became less and less concerned with *Chopin* and became more and more friendly with *Wakasunga* and her child. Just as *Gwekulo* had done before with *Chausiku* and *Chopin* and their one-way relationship, so she served *Wakasunga* and her son, *Linta*. When his mother was unwilling to groom him assiduously enough, *Linta* would plop down with his back to *Gwekulo*, as though he were a sultan, and wallow in her attention.

Mother–child pairs who are the best of pals share core areas or have core areas that are close to each another. We are not sure if they live close to one another and become the best of pals, or whether they get on well and so live in close proximity to each other. Perhaps the latter description fits best. A young, immigrant female goes through hell just to hold down a core area. Over the course of several years, she interacts with many other females, gets in good with one of them, and establishes a core area near this female's.

Social ranking among females is not as clear-cut as it is with males, but it is very much related to how much time you have 'in-group', after immigrating. Thus, the older you are, the higher your rank. As your rank increases among females, you 'earn' the right to select your core area.

In M-group, females of higher rank tend to pick a core area in the middle of the group's range.

However, there is no rule without exceptions. As we observe more and more immigrant females, a particular aspect of personality of females has proven to be an important factor that decides dominance ranking: assertiveness. For example, *Cynthia* dominated many adult females, although she was a small-sized female and her sole offspring was still an infant son. Moreover, contrary to my expectations, those natal females who did not emigrate did not acquire higher status compared with like-aged immigrant females, although the former had belonged to M-group for at least 11 years more. The mechanism that decides dominance ranking among adult females seems to be more complicated than I had thought before.

7

Sexual strategies

In an effort to maximise their reproductive output, males and females may cooperate when it is to their mutual advantage. However, they may also control, manipulate, betray, and even attack each other when they have conflicting interests. Chimpanzees maintain a promiscuous mating system, for which females develop a huge swelling of sexual skin and males large scrota. These morphological features make sperm competition the characteristic feature of chimpanzee sexual activity. Males are expected to mate with as many females as possible, while females are expected to be more choosey because reproductive output over their lifetime is more limited. A powerful constraint on female chimpanzees is the need to consider males' tendencies toward infanticide.

7.1 INITIATION OF MATING

7.1.1 Which sex takes the initiative in mating?

'Copulation is almost always preceded by a male courtship display that signals the sexual arousal of the male and attracts the attention of the female', said Goodall on the Gombe chimpanzees (Goodall 1986). However, our findings show that males take the lead only 46–80 per cent of the time, with females otherwise taking the lead (Nishida 1997a, 1980). Furthermore, the data from Yukio Takahata's 1981 study (Takahata *et al.* 1996) reveal that females in M-group take the initiative in terms of 'approach', 'approach within 3 m of the other', and 'leaving' in addition to 'solicitation of copulation' (courtship). Why do such remarkable differences exist across populations?

The different results between Gombe and Mahale might arise from differences in the definition of 'initiative': an approach to an individual of the opposite sex is sometimes difficult to confirm in the natural environment. At Mahale, initiation was operationally defined as beginning after either a male or female approaches near enough to notice the presence of its counterpart individual by sight or sound. Then the individual who first engages in a courtship display is defined as the one taking the initiative. It seems that a similar method was used at Gombe.

The differences between sites actually may be due to demographic differences in the numbers in age-cohorts of females in the groups studied. Demographic influences may work as follows: as adolescent and young nulliparous adult females are not attractive as mating partners to adult males, these females must more assertively pursue males and thus assume a more active stance for mating. Consequently, they attempt more courtship displays than do parous females. There were probably more nulliparous females in M-group than in the Kasakela group of Gombe, and this difference may well have influenced the proportion of female-initiated courtship display.

7.1.2 Male courtship gestures

The above considerations aside, penile erection is the *sine qua non* of copulation, and thus it is reasonable to expect males, rather than females, to take the initiative in mating and, consequently, to use a more diversified repertoire of courtship gestures. Moreover, as oestrous females are many fewer in number than adult or adolescent males, the females are advantaged by their higher ‘scarcity value’. This is shown by the differences in female greeting behaviour in two reproductive conditions: adult females must pant-grunt to all of the adult males and senior adolescent males. However, when she enters oestrus, she seems to realise her attractiveness. In this state, she may neglect even to greet adult males! For as long as I have studied cycling females, they have pant-grunted less often to adult males during oestrus than during non-oestrous periods. One female never pant-grunted to any male while she was in oestrus.

As already mentioned in Chapter 5, the leaf-clipping display and shrub-bend display are trademark Mahale courtship displays. Females make very few pre-intromission gestures, but most of the time only gaze and present (Tutin & McGinnis 1981; Nishida *et al.* 2010).

Depending on the male in question, each individual has various ways of sending the amorous message, but the female gets the true picture from noting the most basic signal: the erect penis. Females who are willing to copulate may jump into position after simply catching sight of an adducting erect penis. Thus, in any pattern of courtship display, the male has to present his erect penis in plain sight for the female to perceive his intentions. In response to this male courtship display, the female presents her buttocks – that is, she raises her rump to meet the male's penis.

So, why is a male's courtship display not strictly straightforward? Such complication occurs because without a female's cooperation, copulation could not take place. As males on average are larger in body size than females, they theoretically could force females to mate. An older adolescent male makes every effort to get every female to pant-grunt to him (Chapter 3). He does this in order to coerce mating with a female, so long as they are alone. However, because there are usually many males around, if a female does not want to copulate with a certain male, all she has to do is release an ear-piercing shriek. In an instant, a higher-ranking male will intervene and drive away the other males, rescuing the female from having to copulate with the lower-ranking male. Even an alpha male cannot always forcibly mate with a popular oestrous female, as subordinate males may join forces to repel his advances. Copulation is a form of trade, and whether or not it is necessary for females to advertise is merely a question of supply and demand. Females with sex appeal have no need to publicise this, but those females who are not attractive to males must behave in a more assertive manner.

7.1.3 Female courtship displays

A female's courtship signalling is rather straightforward: she presents herself in front of the male and assumes the crouching position (Fig. 7.1). This posture is nothing more or less than an 'intention movement' when it comes to mating. Like penile erection in males, swollen sexual skin is enough of a signal to show a female's willingness to mate. Some female courtship tactics are not so straightforward, such as the leaf-clipping display described earlier (Fig. 7.2). An idiosyncratic pattern of courtship display – a kind of sex dance – was performed by an elderly female, *Wakilufya*. She stood bipedal before a male, raised one arm, and shook a branch or hopped about like a rabbit. A similar courtship display is seen among juvenile chimpanzees at Mahale, adolescent female chimpanzees at Arnhem Zoological Park (de Waal 1982), and even, it seems, bonobos (Kano 1998). The sex dance appears to be a ritualised behavioural pattern of sexual frustration.

Fig. 7.1 An adult male copulates with an oestrous female.

Fig. 7.2 An adult female, *Gwekulo*, performs the leaf-clipping display.

7.1.4 Genital inspection: sexual skin appraisal

Complementing the behaviour of courtship displays by oestrous females, adult males typically perform a 'genital inspection' of anoestrous females (Fig. 7.3). When a male encounters a female, he routinely inserts his index finger into her vagina and then gives his finger a sniff. As males and females spend most of their time apart, by conducting this check the male can probably discern when she will resume oestrus.

Fig. 7.3 An adult male inspects the genitalia of an adult female.

In investigating within which part of the female's oestrous cycle the male most often conducts these sexual skin appraisals, I found that they are done most frequently in the final few days before oestrus (de Waal 1982). As discovered many years ago, primate vaginal pheromones are made up of fatty acids (Michael & Keverne 1970).

7.2 SEXUAL SELECTION

7.2.1 Chimpanzees quick mating

A typical mating by chimpanzees at Mahale proceeds as follows. With his hair standing on end, the male sits on the ground and spreads his legs wide open to exhibit his erect penis. He gets the attention of the oestrous female by plucking a leaf and pulling it between his lips, creating a ripping sound. With hair still bristled, he squares his shoulders and slightly raises one arm. The female then runs up to the male, presses her backside against the male's crotch, and crouches. The male then grasps her hips, inserts his penis, and thrusts at high speed. After about seven seconds he ejaculates, when he may or may not let out a muffled pant. The female lets out a raspy shriek, darts a few metres away, and then crouches again. After this sequence of events, the male often approaches the female and begins to groom the hair on her back (Nishida 1997a). The sequence is similar at Gombe and Mahale, but leaf-clipping courtship occurs only at Mahale.

Chimpanzees are the closest living relatives of humans, but their sexual behaviour is so different that it is shocking. First, chimpanzees typically copulate in public, with everybody watching. Sometimes copulation is done in private, but this 'consortship' occurs only when a non-alpha male does not want a higher-ranking male to intrude during mating.

Furthermore, the time spent in intromission is very short. For ten of M-group's adult males, the average copulation was 7.1 seconds. Similarly, a 7.0 second average was recorded at Gombe (Tutin & McGinnis 1981). It is even more amazing that individual differences are miniscule, with everyone in M-group taking 6–8 seconds. Age is not a factor in mating duration, except for juvenile or infant males, who may extend their mating for more than 20 seconds but without the capability of ejaculation. Males in their late thirties do not mate longer than males in their late teens. There is no foreplay; instead, sperm competition is severe among male chimpanzees, with their huge scrota (Short 1979). Only males who quickly finish ejaculation can inseminate females before being interrupted in their copulations. Foreplay would immediately invite robust interference. The apes' context differs greatly from humans, who mate in private, typically without competitors.

The extended female orgasm is thought to be unique to human sexual behaviour. However, the physiological phenomena involved in orgasm, such as increases in vocalisation, body heat, and blood pressure, gained recognition as being authentic in studies of macaque monkeys in the 1970s (see Goldfoot *et al.* 1980 for orgasm in macaques), so it is likely that such phenomena also occur with the female chimpanzee. Nevertheless, as the female chimpanzee runs away ('darts') from the male after he ejaculates, there seems not to be the distinctive phase found in human female orgasm, such as the female feeling faint after she climaxes. Why does the female chimpanzee run from the male immediately after copulation? I speculate that the female is seeking to avoid possible assault by other males, which might occur if she were to continue crouching in front of her current partner. If this is the case, why doesn't she keep silent at the end of copulation?

It has been suggested that female copulatory vocalisations advertise the position of a copulating pair to other group members, thus serving to incite inter-male competition (Hamilton & Arrowood 1978; Hauser 1990). This hypothesis seems not to hold for chimpanzees. First, why would they emit such a call at the *end* of copulation? Wouldn't this be too late to incite competition? More to the point, I have never observed a female's copulatory squeals attracting other males to the scene or provoking inter-male competition. Hasegawa and Hiraiwa-Hasegawa (1983, 1990) also reported that copulating pairs rarely evoked aggressive responses from adult males and, furthermore, that younger females emitted copulatory squeaks more often than older females. Dixson (1998) advanced the hypothesis that these vocalisations are made to alert other males in the group to the fact that mating is in progress, thus encouraging them to approach or attempt mating at some *later time* (his italics). I cannot test this hypothesis because continuous following of oestrous females is extremely difficult in the thick bush of Mahale. My hypothesis is that the female vocalises to express her satisfaction to the partner, which might prevent the male from engaging in infanticide after the female mates with other males later.

The copulating position is almost always dorso-ventral. When adolescent females copulate with juvenile males in trees, the frontal position may occasionally be used, but this is never seen between mature males and females. Dorsal mounting may be convenient for males, who must finish insemination as quickly as possible before other males come to interfere. On the other hand, frontal or ventro-ventral mating is common in orangutans and gorillas.[1] However, these great apes do not mate in the presence of other adult males, so they are not under pressure to finish quickly. The problem with this line of conjecture is the bonobo, who, like the chimpanzee, engages in promiscuous mating, but, in stark contrast, exhibits frontal mating about 30 per cent of the time. This riddle still eludes my attempts to solve it. Maybe sexual interference by other adult males is rarer among bonobos than among chimpanzees. Moreover, the mating duration of the bonobo is somewhat longer than that of the chimpanzee, about 20 seconds or so (Kano 1992). This also suggests that direct inter-male competition is less frequent among bonobos.

The long penis and large scrotum of males and the huge sexual swelling of oestrous females in chimpanzees and bonobos are explained by the severe sperm competition inside the female reproductive organ, which results from their promiscuous mating systems (Short 1979).

Chimpanzee promiscuity may explain the short mating duration, large testicles, slender penis without glans penis, large female

[1] See Nadler (1995) for orangutan sexual behaviour; see Nadler (1976) for gorilla sexual behaviour.

sexual skin, and dorso-ventral copulatory posture. On the other hand, the fact that human mating is basically monogamous, although often sequential, may be related to long duration of mating, mating in private, thick penis with glans penis, small testicles, deep female orgasm, and customary ventro-ventral mating posture.

7.2.2 Female sexual cycles and female popularity

Copulation takes place at any time on any given day, but it occurs more often in the morning than in the afternoon. The early morning is an especially prime time for mating (Hasegawa & Hiraiwa-Hasegawa 1983, 1990). Chimpanzees do roam at night, and I confirmed this with M-group once, by hearing a vocalisation from a large party on the move. Kohsei Izawa observed a big party travelling in the Kasakati Basin at about 01:00 a.m. (Izawa 1975), so it is possible that copulating goes on at night too.

Nevertheless, once past juvenility, both males and females make and sleep in separate beds. Night-time activity raises the danger of predators. Therefore, even if we conjecture that chimpanzees copulate at night, the idea that night-time copulation takes precedence over day-time copulation seems highly implausible.

Cycling females show swelling of their sexual skin for 7–14 days out of the 35–40-day menstrual cycle. During this period, the swelling publicises their availability for copulation. Chimpanzees are promiscuous, and so no individual perennially retains fidelity to one particular mating partner.

Chimpanzees have no permanent sex partners. It is hard to imagine a male and his consort living together permanently, as females do not copulate for 4–5-year periods, from late pregnancy to the infant's weaning. This line of thinking appears reasonable, but as gibbons maintain monogamous relations for several years even without having sex, such pair-bonding also could, in principle, occur in chimpanzees.

In addition to being in oestrus for only about ten days every menstrual cycle, after weaning her child and resuming oestrus, a female gets pregnant within a few months. Mature females, provided they are not barren or afflicted by the death of their infant, are almost always either nursing or pregnant following their adolescent years and thus have only about 200 days over their lifetime for copulation. Because a male's direct support is not necessary for child-rearing, after he has impregnated one female, by moving on to the next oestrous female, he can maximise the production of his own progeny.

Males play the 'promiscuity game' because, for them, it is clearly more advantageous, but this is not the case for females. Even if a female copulates with many males, this does not ensure that she will give birth to more infants. Accordingly, females aim to pick and choose their male partners, with a strong inclination not to allow copulation too readily. This general rule applies to all creatures, from fruit fly to human, not just the chimpanzee (Hasegawa & Hiraiwa-Hasegawa 1983, 1990).

However, female chimpanzees in the early stages of oestrus, when there is no chance of conception, seldom are choosy about their male. The same female, compared between the early and late stages of oestrus, sees her popularity with the males change significantly. In the early stages of the oestrous cycle, she tends not to show mate choice. Although *Silafu* had a three-year-old daughter, her oestrous returned while I was tracking her. Before daylight she arose from her bed, hit the ground running, climbed another tree where a male was still asleep, and hopped in bed with him, prodding him with her buttocks, apparently begging for copulation. *Silafu* kept this up for about half an hour, eventually copulating with seven males. However, in the latter stage of her cycle, she stopped sending such invitations to males. The benefit of promiscuity for females may be increased confusion over paternity, as adult males are known to kill and eat small infants (see Chapter 6).

The last few days of a fully swollen sexual skin is when a female becomes most irresistible to a male. The day of ovulation is just before the stage when the sexual skin's swelling begins to decrease (Tutin & McGinnis 1981). As the possibility of pregnancy rises, the alpha male begins displaying possessive behaviour (Hasegawa & Hiraiwa-Hasegawa 1983, 1990). He may accompany her the whole day, and if another male so much as approaches, the alpha male will chase him away in order to monopolise the female. The female over whom the strong alpha male has taken charge copulates only a few times the entire day (Nishida, unpublished data), whereas a 'free' female in oestrus may copulate 20–30 times with several males. In the evening, the alpha male makes his bed below that of his consort, in order to keep watch on her throughout the night (Nishida 1981). This suggests that an oestrous female occasionally slips away from the alpha male at night to seek a male she prefers. However, this is hard to do because chimpanzees sleep supine and the alpha male can directly watch the female from below if he keeps his eyes open.

Since sperm can survive in the female's reproductive tract for a couple of days, her ovum attracts the sperm of all of these mating males, which becomes the crucial 'survival of the fittest' race called 'sperm competition'. Even if he cannot monopolise females, the male who produces the most massive quantities of sperm and penetrates female after female will leave the most children behind. For this reason, male chimpanzees have testicles of colossal proportions (Short 1979).

7.3 MALE AND FEMALE SEX APPEAL

7.3.1 Male sex appeal

A spectacle I never will forget was the gesture of an adolescent female named *Lyinso*, who had just immigrated into M-group. It seemed that she had fallen in love with *Aji*, a young adult male. She had this wonderful way of coming up close behind *Aji*, standing bipedal, and brushing his shoulders ever so slightly with one hand, then peering into his eyes. She was terrified of us, and after hiding in some bushes, she crept out and repeated the gesture to *Aji*. He snubbed her, ignoring her advances.

Social rank does not necessarily determine a male's popularity with the females. *Bembe* was a male of large build but who was surpassed in rank by males that were his junior. Despite his lowly status, he was attractive (amazingly!) to some females. If observations were confined only to copulations amidst scenes swarming with males, then documenting *Bembe*'s behaviour would be impossible. But if you tracked *Bembe* in private, you would usually find him romping about alone, then soon after just sitting and staring into the undergrowth. Out of nowhere, an oestrous female would appear on the scene and *Bembe*, cool as a cucumber, would start copulation.

What kind of male do female chimpanzees fancy? The fussiness a female has over her favoured type of male is especially apparent when she refuses to copulate. In the early 1990s, there were nine adult male members of M-group: *Bakali*, *Musa*, and *Ntologi* were in their late thirties and so were the older members of the group; *Kalunde* was about 29 and middle-aged; then there were *Bembe*, *Nsaba*, *Aji*, *Jilba*, and *Toshibo*, who were all young adult males in their late teens. The males who made advances but were often rejected by oestrous females were two low-ranking older males, *Musa* and *Bakali*.

In 1992 I witnessed about ten incidents when an oestrous female, after an adult male merely approached her, let out a little shriek and ran off in the direction of another male. A mere encounter with a female could bring the sting of rejection. *Musa* met with this bitter fate seven different times, and *Bakali* twice. Later that year, instances of an oestrous female letting out a little shriek and fleeing the courtship scene happened 18 times in the same season. This time, *Bakali* faced this kind of rejection 11 times, and *Musa* 6 times. By reviewing the observations of these two old males, we find that four females refused each of them. After being rejected many times and still coming back for more, both *Musa* and *Bakali* resorted to force, actually kicking and punching uninterested females. But before long the knights-in-shining-armour of M-group, alpha male *Ntologi* and young macho males such as *Aji*, came to the rescue of the damsels-in-distress.

That year, *Musa* especially got himself into trouble with a middle-aged female, *Zip*. Although *Zip* would not mate with *Musa*, he followed her like a bee drawn to honey; when the group was on the move in a procession, he walked just in front of her, menacingly. *Zip* could not stand this, so she let out a shriek of disgust. *Musa* then stopped and *Zip* overtook him in the line. *Musa* kept up this kind of 'stalking' for several days.

The alpha male, *Ntologi*, who was similar in age to the two other seniors (*Musa* and *Bakali*), was never rejected when he solicited copulation. One must acknowledge that no matter how revolting a female finds the alpha male, she cannot reject him. If a female rejects a lower-ranking male by letting out an ear-piercing shriek, a male higher in rank typically will come to her aid and chase away the other male. However, no single male can chase off the alpha male. Therefore, it is impossible to know whether the alpha male avoids rejection because his status makes him the fancied one or because the female cannot resort to other males, even if she does not like the alpha male. Only when higher-ranking males join forces can the alpha male be prevented from copulating.

On occasions when many adult males, including the alpha male, are competing fiercely for a charming oestrous female, any adult male might not succeed in mating with the oestrous female for a prolonged period of the day. However, when one male achieves mating with her, most of the other males do so, one after another.

Taking a more assertive stance, a female may solicit a male for copulation. This is called 'proceptive' behaviour. What I call the 'erection check' is one example of this kind of behaviour (Stumpf & Boesch 2006). The oestrous female slouches about, peering at the male's lower quarters to see if his penis is erect or not. I am convinced that she checks only the males she wants to mate with. Why else would she look, if she does not want to copulate with him? In 1992 females were seen to make ten erection checks. The recipients were five adult males, four of whom were among the group's five young adult males in their late teens. These four younger males were checked a total of nine times, the other examined male being middle-aged. However, the three old males of the group (including the alpha, *Ntologi*) never had this examination performed on them by the females. Moreover, in 1981 an erection check was shown by a prime adult female, *Chausiku*, who had a weaning infant, to a young adult male when there were only two adult males left in K-group. This was proceptive behaviour by a female who found it difficult to obtain an alternative male. When *Chausiku* found her mate's penis to be flaccid, she even stimulated it to erection by repeatedly rubbing it lengthwise with one hand (Nishida 1981).

In all, females performed 17 courtship displays. Four of the five young adult males were invited to copulate. One old male was also solicited, but it was *Ntologi*, M-group's alpha male.

Likewise, at Taï, female promiscuity depends on the stage of their sexual cycle. Resistance and receptivity to male solicitations were examined. Many females exhibited preferences for particular males, and female proceptivity rates were lower and resistance rates were higher in the peri-ovulatory period, during which unwanted copulations were averted (Stumpf & Boesch 2006).

7.3.2 Female sex appeal

Adolescent females show their first maximal sexual swelling at around the age of 10 years, but they do not attract sexually mature males. Adolescent females who have just immigrated into the group are no exception. Therefore, their lack of popularity seems to be due not only to their being natal females but also to their young age. With further progress made in research, we have confirmed that birth experience is the most important factor contributing to the female's attractiveness: adult and adolescent male chimpanzees prefer parous to nulliparous females. Therefore, a nulliparous female will walk about endlessly looking for the willing (or unwilling) juvenile or infant male, thrusting her buttocks in the air, or even for the infant males who can manage to attain an erection – seeing them penetrate a female for the first time is somewhat startling to an observer (Fig. 7.4).

Gwekulo was a middle-aged but barren female and thus nulliparous. Accordingly, she became an 'unwanted' and frustrated female, and

Fig. 7.4 An infant male mates with a young oestrous female.

she approached any male regardless of age, made an erection check, and then tried to be intimate with him. If he was not erect, she became enraged and begin attacking the shrubs around her.

An extraordinary distinction between human and chimpanzee is that our standard for what constitutes a preferred female seems to be rather different. For human males, an extraordinarily attractive female is almost always a pretty young woman. Chimpanzees typically find nothing appealing about a young, nulliparous female. Rather, a parous, somewhat weathered female with an infant just weaned seems most appealing. Even if she is 40 years old, adult males go crazy for the oestrous old female and engage in such intense competition for the female that they forget even to eat! If one of the males manages to mate with her in public, the other males become angry and bark at him, even if the successful male is the alpha. Consequently, the male chimpanzee's lust is straightforward, simply focusing on reproductive success in seeking females who are veteran mothers, able to give birth to another offspring.

Unlike chimpanzee males, human males invest much in caring for their offspring, and thus female chastity is considered most important for their reproductive success, making virgins and young women greatly valued. Although I have stressed the various differences, it is quite possible that human males also have chimpanzee-like tendencies to some extent when it comes to taste in females. There is a Japanese proverb: 'Stealing is best, maid second', meaning stealing another male's wife is best, mating with unmarried women second, and mating

Fig. 7.5 Oestrous females are often attacked and bitten on their sexual skin by adult males.

with your own wife last. Perhaps humans merely repress their carnal hungers, pushing them into a corner of the unconscious mind.

Adult chimpanzee males make a fuss over a female who has just weaned her infant and whose oestrus has resumed, regardless of her age. When such a female is around, the male's eyes sparkle, he skips meals, and goes head-over-heels for her. Her presence puts a hold on all copulating, with endless quarrels over her, day-in and day-out. Not only do rival males come away with battle scars, but this one-of-a-kind female gets bitten on her sexual skin in all the commotion (Fig. 7.5). Such turmoil can even trigger a power struggle that results in the ascendency of a new alpha male (Nishida 1983a) (see Chapter 8).

The great paradox is that, despite all of the interference and conflict, this highly desired female actually ends up copulating less often than does the most unattractive female of the group. Moreover, a strong alpha male subsequently shows 'possessive behaviour' (Tutin & McGinnis 1981) towards the female in the peri-ovulatory period, preventing any other male from mating with her. I once followed the alpha male, *Ntologi*, and his consort for a whole day, and he mated with her only twice. Therefore, the popularity of a female chimpanzee cannot be measured simply by the number of times she mates. Among chimpanzees, it appears that the capacity for childbirth and rearing is a critical factor in determining one's sex appeal, and thus sexual sprees can occur even with females who are in their early forties. As already stated, even females over 40 years of age give birth (see Chapter 3), although from that age, they have so far failed to raise their children to weaning age.

7.3.3 Chimpanzee elopement

A kind of elopement occurs between chimpanzees, in which a non-alpha male will beckon a female to slip away with him. This has been labelled 'consortship' (McGinnis 1979) or 'safari behaviour' (Tutin 1975). From outside, it resembles the honeymoon of newly weds (Fig. 7.6). When the male does not want to be bothered by other males, he coerces his partner to travel with him to the edge of the range. The female typically resists his travel-bound proposal. If she refuses, the male must spend much time patiently pleading until he succeeds. For example, whenever the male begins to walk, she may stop after following him only 5 m. Only when she finally gives up because no one comes to help her, or the male's aggression is too strong, will the pair spend the next few days, weeks, or months together avoiding the group. The longest safari I have recorded thus far lasted three months. Since a female lets out a rousing shriek if she declines an invitation to safari, safari behaviour cannot arise without some degree of the female's cooperation. At the very least the male must be popular enough to gain the female's consent to embark upon a safari.

In the case of a large group, such as M-group, which had 30 female members, at any given time a few females will be in oestrus,

Fig. 7.6 *Kamemanfu* and *Wakapala* in consortship.

so to a certain extent males can also afford to be choosy. In my experience, it is the second-ranking male who often takes a female on safari, perhaps because it is easier for him than for subordinate males to coerce her to accompany him, since only the alpha male could interfere. An alpha male can show possessive behaviour in front of other males and thus does not need to sequester himself with his consort.

7.3.4 Who left many offspring?

What kind of male fathers many children? In the early 1980s, as in the case of the Gombe chimpanzees, Hasegawa and Hiraiwa-Hasegawa clarified that the alpha male showed possessive behaviour towards oestrous females in the peri-ovulatory period (Hasegawa & Hiraiwa-Hasegawa 1983, 1990). Therefore, he monopolises a fertile female if he wishes to do so.

Recently, in the early 2000s, Eiji Inoue did DNA paternity testing for sexually mature males. He clarified that of 20 paternal candidates, the fathers of 5 of 11 infants were 2 alpha males (Inoue *et al.* 2008). Moreover, genetic relatedness among adult males is significantly closer than among adult females. This high degree of relatedness is expected from male philopatry, but the fact that a particular male, such as the alpha male, produces disproportionately many offspring strengthens this tendency.

The results reinforced what we had long expected. If high-ranking males did not produce more offspring than lower-ranking ones, why would they strive for rank? Only the production of many more offspring is what could have fixed in evolution the genetic tendency to seek higher status.

8

Male political strategies

Male chimpanzees compete for social status. Winning the top rank in the dominance hierarchy appears to be the main lifetime goal of most, if not all, male chimpanzees. They breathe, eat, and live to dominate others. Why are they so driven to become the dominant ape? Because higher rank confers higher reproductive advantage through more frequent mating opportunities. Alpha rank, in particular, confers a disproportionately higher chance of mating. To gain promotion in dominance rank, a male must perform various activities at different stages of his life: pant-grunt to adult males before maturity; play-fight and wrestle during juvenility; harass adult females throughout juvenility and adolescence; establish dominance status among adolescent male peers; and perform intimidation displays, including 'display contests', from late adolescence into adulthood. The last requirement is a lifetime activity for all adult males.

8.1 MALE INTIMIDATION DISPLAYS

8.1.1 Set up, charge, and climax

Adult male chimpanzees have a repertoire of 'displays' that are noteworthy behavioural patterns, to say the least! The charging display, so named by Jane Goodall (1968), involves many behavioural elements, some of which involve the use of tools. Charging displays have three stages: set up, lunge and dash, and climax.

In the 'set up' stage, the male bristles and violently shakes a shrub or vine with his hands, making a commotion. During the dry season, piled-up dead leaves provide an additional prop for the male charging display. In it, the male begins to 'scrub', that is, rake and stir the pile of leaves with his hands.

In the 'lunge and dash' stage, he runs with thudding sounds by pounding and stamping on the ground, or dashes around while slapping the ground with a hand. Or he drags a stick behind him, bending or breaking trees as he rushes by. Sometimes, he just dashes about. If he is initially sitting and scrubbing dry leaves, then he begins to run, clutching a bunch of dead leaves in one hand.

In the 'climax' stage, if he drags a stick behind him, he tosses it into the air for his finale. If he carries a pile of dead leaves, he finally strews them all over the place, like confetti. At other times, large rocks or tiny stones are thrown. Kicking and beating a buttress root is another favourite and frequent finale.

There are two types of display: one is accompanied by pant-hoot vocalisation from the onset; the other is silent. The former's climax ends in grandiose 'whooping' vocalisations. Vocal display is propaganda and typically is not aimed at particular individuals. On the other hand, silent display is intimidation and mostly aimed at particular individuals, although the performer does not actually hit them.

On other occasions the displayer stands up and, with alternating feet, drags the dry leaves while making a huge din, sometimes finally breaking out into a mad dash.

A simplified version of the charging display does not entail a progression of all three stages. Moreover, some behaviours are unrelated to the charging display. For example, the male may merely stamp his feet in response to a vocalisation from afar. On the other hand, in a lengthy display or one that lasts several minutes, it can be hard to identify which part is the climax.

If you tag along behind a male, you notice him letting out small pants in a kind of 'hoooo, hoooo' voice. Eventually his gait speeds up, and finally he bursts into a dash and kicks a buttress root with a thud! As there are also cases of pant-hoots starting even 100 m before reaching the root, this is not in response to the stimuli of seeing the buttress root, but rather the chimpanzee becoming aware that he is approaching the buttress root's site, and without a doubt, a charging display will ensue. It appears that the males have memorised each and every location of drumming spots within their range (i.e. the topography of buttress roots). I realised this from observing that adult males typically kick and slap the buttresses of the same trees along the trails.

In order to show-off ostentatiously in front of many group members, adult male chimpanzees often avail themselves of streambeds where suitable missiles are abundant. They run rampant, dragging branches

Fig. 8.1 A charging display – branch-dragging.

along the ground, beating the ground with sticks, and pulling and shaking woody vines (Fig. 8.1), as well as rolling rocks by pushing or dragging them with one hand and throwing sticks and stones. Huge rocks are often deliberately heaved at the water's surface with both hands in a kind of splashing display (Nishida 1993b), but rocks are not thrown at rivals in direct confrontation.

Before performing a charging display, the male must be extremely cautious and ensure that a higher-ranking male is not in the vicinity. It is common for a young adult male, midway through his display, to become scared stiff and immediately stop his performance, realising that all this time a high-ranking male was nearby in the bushes. When a display is performed near a high-ranking male, he appears to interpret it as a challenge to fight and thus follows up with an assault. Therefore, displays are restrained anywhere near a powerful male, which is one reason why high-ranking males exhibit such displays much more often than do low-ranking males.

However, an ambitious young adult male sometimes fails to notice the proximity of a dominant male. On such an occasion, the young adult male is usually punished by the dominant. However, the dominant male sometimes ignores the young male's bad manners. It may be that he does not consider such apparent disrespect as a challenge against him – that is, that he distinguishes an intentional challenging display from an unintentional one. Like humans, chimpanzees seem to differentiate intentional from unintentional behaviour. Let me illustrate this with an example.

Once a party of chimpanzees was descending from a large food tree, and an adult male, *Kalunde*, was hit on the side by a stool falling from an adolescent male overhead. *Kalunde* glanced at the adolescent male and simply wiped the faeces off with a leaf, without confronting the younger chimpanzee. I have witnessed many instances where an adult chimpanzee gets splashed with urine from an infant overhead, and in every case the adult simply glanced at the infant and did no more. It appears that they understood the infant's innocence. At Gombe, the subordinate male *Mustard*, displaying above ground, fell from the crown of the tree onto the back of the alpha male, *Goblin*. Since the alpha male did not attack *Mustard*, we can conclude that he 'seemed to take it well'.

8.1.2 Display of marked individuality

Whatever display is performed depends on the personality of the male. For example, in August 1992, when *Ntologi* and *Kalunde* (two high-ranking adult males) entered the valley, it was taken for granted that both of them would toss a heavy rock into the water below. Spectators of this display are overwhelmed by the splash and sound alone. When tracking M-group in 1992, if you heard this great explosive sound you could guess (at a probability of 70 per cent) that it was either *Ntologi* or *Kalunde* displaying bravado (Nishida 1993b).

At the climax of his pant-hoot display, *Nsaba* stomped his feet, whereas his peer, *Aji*, slapped the ground with one hand. *Fanana* also slapped the ground with one hand, a habit that continued since his adolescence. As *Shike* dashed about, he repeated his rhythmic stomping and slapping of the ground; this display was so amazing that the Tongwe people gave him the nickname *Bwana Bujege*. 'Bujege' is a traditional ceremony the Tongwe people hold to drive out the elephant demon, in which a dancer stamps the ground with great fervour. In other words, *Shike* was called a master *Bujege* dancer. It got to the point that even when we were far off in the distance, we could still recognise *Shike* from the brilliant sounds of his dance.

The volume of display repertoire or duration of display varies from male to male. In his heyday, *Ntologi*'s display often dragged on for more than several minutes, in contrast to the maximum 20 second duration of ordinary adult males, but as he approached old age, it shortened. To balance everything out, he performed the shortened display at a much higher frequency.

Our lodgings were tucked into a small valley called Kansyana. For about ten years, from 1975 to 1985, we used a kind of aluminium housing facility made by Uniport as part of our living space. Soon the sheet-metal walls of the Uniport structure became a prop for the chimpanzees' displays. They recalled the racket-making capability of metals from their experience of drumming on the corrugated iron door of a small storehouse for bananas and sugarcane, which we used when we were still feeding the chimpanzees. They used to kick and slap the door when they discovered there was food within. The storehouse was only 50 m away from the Uniport house.

Since the Uniport house was in a well-lit spot, we were able to do some really detailed work as they put on some amazing performances. In this way, the character they assumed in their displays became very clearly defined (Nishida 1994, 2003b).

Females and most of the male youngsters never slapped, hit, touched, or kicked the metal wall, as if they were afraid that touching it would bring the worst misfortune. The only exceptions were: a juvenile male, *Chopin*, hit the wall lightly with his right fist at age seven. An adolescent male, *Masudi*, once 'kissed' the wall when he was travelling past the house at the age of 13 years. An 11-year-old male, *Bonobo*, stopped and lightly put his right knuckles on the wall, which naturally did not make any noticeable sound. Adolescent males typically did not touch the wall, but all adult males sooner or later began to slap or kick the metal wall. Therefore, this metal-drumming seems to have become the 'initiation' ceremony for adult males: it became a local culture of M-group! However, the exact age at which males began to hit the metal wall varied.

All of the other adult males hit the aluminium wall with hands flat-palmed, but *Bembe* hit it with a limp fist, producing a faint knocking sound. *Nsaba* stood bipedal and, using only his left hand, slowly hit the wall several times (Fig. 8.2). *Kalunde* typically approached at high speed, stood bipedal and beat the wall with alternating hands, then sped off. His display packed an incredible punch (Fig. 8.3), so I could tell he had arrived, just from his drumming sounds. *Musa* usually pointed his backside at the wall and, standing on all fours, let fly a barrage of short, quick kicks with his left foot (Fig. 8.4).

Without a doubt, the highlight was the alpha male, *Ntologi*. First, he stood bipedal and hit the wall with his right hand (Fig. 8.5), then he got on all fours and, with either his left foot or both feet, hit the wall

Fig. 8.2 *Nsaba*'s slapping display.

Fig. 8.3 *Kalunde*'s slapping display.

Fig. 8.4 *Musa*'s kicking display.

Fig. 8.5 *Ntologi*'s slapping display.

with a flying kick (Fig. 8.6). Thus, he used both feet and hands in his display. *Ntologi* was also unique because he hit not only the Uniport structure but also another prefabricated storehouse that was built later.

The apes knew the tightness of their drum and to a large extent were creative in its use. Regardless of the male, in the case of the metal wall, hands were used significantly more often to hit the wall versus

Fig. 8.6 *Ntologi*'s kicking display.

the display of using the feet to kick buttress roots. To explain this difference, we conjectured that pounding on the aluminium by hand produced a sound considered to be loud enough, making the greater force of kicking unnecessary (Nishida 2003b).

Interestingly, adult male chimpanzees sometimes came to the camp only to hit the metal wall. For example, one day when a large party passed about 500 m west of the camp, going from north to south, an adult male, *Aji*, made a deliberate detour to the camp and hit the metal wall several times before immediately returning to the west to rejoin the party! Another day, around 8 o'clock in the evening, an adult male, *Masudi*, came from the south and hit the wall and emitted a pant-hoot, then continued to travel northward.

I conclude that males perform displays to show-off their strength. The intimidation display is used to demonstrate the male's strength and stamina as a powerful competitor to fellow males and his status as an attractive sex partner to females, without taking the risk of getting wounded. This hypothesis is supported by the fact that high-ranking males do a great deal of displaying, while low-ranking males restrain from doing so. Furthermore, spectators – including adult females – intently watch the displays and emit loud 'hwau' calls at the climax of the display.

Displaying probably is also a way to announce one's current location to an ally (Boesch 1991c; Mitani & Nishida 1993); however, this function is not yet fully understood.

8.1.3 Display contest

Since chimpanzee groups often split up into smaller parties even in the 'gregarious season' (see Chapter 2), even the lowest-ranking male has many opportunities to show-off his strength. When he meets a mother–offspring family group ranging on their own, for example, it presents a good chance to engage in a charging display. He charges at the female and shakes shrubs, swings from woody vines, slaps the ground, and kicks buttresses. Any available items can be used. The female screams as loud as she can, even if she is not wounded. Another male may rush to the scene and begin his charging display at the first male. Then, the first male screams loudly, or at least retreats from the scene. Then, another male appears, and so on. In such a way, I have seen five performers coming one after another. Beautifully, the order of appearance was the ascending order of male rank: the first to arrive was the lowest-ranking and the last one the highest-ranking.

I call such a chain of male displays a 'display contest'. This clearly illustrates how adult males behave towards one another. They not only make every effort to maintain dominance over all adult females, but also nip in the bud the efforts of the lower-ranking males to rise in rank. Display contests are more likely to occur when a prey animal is captured or when someone begins to do a streambed display.

8.2 ALPHA MALES AND POWER TAKEOVERS

The alpha male is the highest-ranking male of a unit-group or community, who never pant-grunts but instead receives pant-grunts from all other group members. Even for a first-timer doing field study on wild chimpanzees, picking out the alpha male is easy, at least at Mahale. As he is the one getting most of the pant-grunts, all you have to do is listen for noisy vocalisations and go to the source. Heading for the scene of commotion among chimpanzees inevitably will bring an encounter with the alpha male, because he assumes the role of authority figure and breaks things up. If another male were to carry on as an authority figure, the alpha male would be livid.

When chimpanzees are ranging, it is never totally clear who is assuming a leadership role. However, the group's alpha male and elder adult males have great influence over the direction and timing of the group's movements. One day *Kasonta*, the alpha male of K-group, sought to set out, but some of the other members could not keep up. Over and over, he shook a tree branch. When he did this, the males who were lying down jumped up to attention, and everyone got off their haunches, following suit.

M-group fording the Lubulungu River (the research site's largest river) is a sight not to be missed. As the ford is a wide part of the river, it might be a high-risk area for lion or raptor attack. Moreover, it is a double-edged sword for youngsters, because the river has a rushing current and the water level fluctuates. The chimpanzees move south in groups of two and three, and upon arriving at the riverbank they typically do not cross immediately but instead pause to rest or to hold a grooming party. Eventually, they form a huddle and, from what I gather through observation, apparently tell each other, 'Okay guys, it is about time to make a move'. Suddenly, *Ntologi*, the alpha male, rises to his feet and begins crossing. Within a few minutes the whole legion crosses the water. Adolescent males may cross first, but other members usually do not try to keep up with them. At times like these, you clearly see *Ntologi*'s operative force. I do not assert that M-group's chimpanzees always wait to cross a river until the alpha male does, but I assert that you can see the strong influence the alpha male exerts on the movement of a large party.

It appears that male chimpanzees make every effort to increase their rank. The paramount merit that comes with being alpha male is the ability to monopolise any oestrous female you desire for copulation (Fig. 8.7). An alpha male may guard his mate all the time,

Fig. 8.7 *Kasonta* with two oestrous females, *Gwekulo* and *Wakilufya*.

preventing other males from stealing his 'wife'. Consequently, he shows 'possessiveness' (Tutin 1979b) towards the ovulating female and thus guarantees the conception of his own progeny. In this way, alpha status confers great reproductive advantage, and the DNA analyses at Gombe (Constable *et al.* 2001), Taï (Boesch *et al.* 2006), and Mahale (Inoue *et al.* 2008) reveal that alpha males, or alpha candidates, sire many offspring. As a secondary advantage, the alpha male has priority rights on nutrient-rich luxury foods such as meat. He not only obtains highly nutritional food more often than other chimpanzees but also gains the chance to use the meat for political gain within the group (see Section 8.9).

8.3 POWER TAKEOVERS IN K-GROUP

8.3.1 *Kasonta*, the tyrant

In K-group, *Kasonta* held the post of alpha male for over ten years, *Sobongo* for two and a half years, and *Kamemanfu* for five years. *Kasonta* was the first alpha male that I got to know really well, but getting to know him took a bit of time. The first unit-group we provisioned, called Kajabala group (K-group), was named after *Kajabala*, whom I had mistaken as the alpha male at the time, although he was actually the second-ranking male.

In those early provisioning days, *Kasonta* never showed himself around the provisioning area. Even when he did decide to come out, it was for an instant to snatch up some sugarcane and then rush back into the bush, so the thought that he was the alpha male never crossed my mind. I observed alone in the earliest years, but he was a coward around humans. However, among his cohorts he was a tyrant, monopolised whichever female was in oestrus, rampaged at will, and had everyone in fearful cries all the time. He also took a kingly share of the bananas and sugarcane that we gave to the chimpanzees.

However, every time the old-aged and large male, *Kasagula*, came on the scene, *Kasonta* hastily wolfed down every last bit of the sugarcane he had just gathered (Fig. 8.8). *Kasagula* grimaced whenever *Kasonta* performed an intimidation display, but why *Kasonta* watched helplessly as *Kasagula* robbed him of his favourite food was the first knotty problem I encountered in my research of the chimpanzees. The relationship I observed between *Kasonta* and *Kasagula* taught me that, in chimpanzee society, dominance is important, but it doesn't determine everything.

Fig. 8.8 *Kasagula* (left) and *Kasonta*.

8.3.2 *Sobongo*'s takeover of alpha position

K-group's males disappeared one after another; the last males left standing were *Kasonta*, more than 35 years old and 57 kg in weight, the youthful *Sobongo*, about 18 years old and 47 kg in weight, and *Kamemanfu*, as old as *Kasonta* and 35 kg in weight.

On 5 March 1976, *Sobongo* was found to have four big wounds on his back, and *Kamemanfu* had a wound on his left hand. It seemed impossible that *Kamemanfu* acting alone could have inflicted such injuries on *Sobongo*, because in 1974 *Kamemanfu* had been surpassed in dominance rank. It was likely that *Kasonta* and *Kamemanfu* united forces against *Sobongo*. For almost 40 days after this incident, we saw no male–male fighting, primarily because the group scattered widely into small parties, and both *Sobongo* and *Kamemanfu* each formed a consortship with a young oestrous female. *Kasonta* led the largest party, comprising three oestrous females and two adolescent females, and ranged along the northeastern boundary of the territory in an area overlapping B-group's territory.

The next time I observed the three adult males together was 17 April. Now the real rivalry emerged clearly. In the early morning, the dark forest was full of the stench of chimpanzee faeces. I could not believe my eyes. *Kasonta* was limping, with eight severe lacerations, on the right ear, head, left arm, both hands, and left foot. I had never seen him with wounds over the ten years since I first identified him in 1966.

Fig. 8.9 *Sobongo* threatens *Kasonta* in a tree.

I thought that no one would dare to challenge this huge, aggressive male. *Sobongo* also had lacerations on the back, face, nape, left hand, and both wrists. *Kamemanfu* had no wounds at all this time. Obviously, *Sobongo* had challenged *Kasonta*'s rule.

When *Sobongo* made challenges, he climbed a tall tree, approached *Kasonta* from above, and threatened him by pant-hooting continuously (Fig. 8.9). This provocation immediately stimulated *Kasonta* to chase *Sobongo* through the trees, but he rarely caught him, as *Sobongo* was much younger and more lightly built (Figs. 8.10, 8.11). Therefore, instead of direct chasing, *Kasonta* more often showed charging displays on the ground (Fig. 8.12). Once, when *Sobongo* was caught by *Kasonta*, they grappled fiercely in a tree until both fell to the ground without obvious injury.

Kasonta was not defeated quickly, because he was supported by the third-ranking *Kamemanfu*. The latter did not often directly intervene in the conflict but typically expressed his support of *Kasonta* by

Fig. 8.10 *Kasonta* watches *Sobongo* overhead.

synchronising pant-hooting and charging displays. In particular, communal pant-hooting seemed to encourage *Kasonta* to defeat *Sobongo* (Fig. 8.13). *Kasonta* and *Kamemanfu* also often embraced and kissed each other with open mouths.

However, this apparently coalitionary relationship did not last long, as I found *Kamemanfu* grooming clandestinely with *Sobongo* in the bush (Fig. 8.14). At that point, *Kamemanfu* changed his coalition partner in a bewildering manner (Fig. 8.15). This attitude of *Kamemanfu*, which I called the 'allegiance fickleness strategy' (Nishida 1983a), had its rationale. At that time, there was a charming middle-aged female, *Wasalamba*, who had just weaned her infant and resumed oestrus (Fig. 8.16). *Kamemanfu* could not mate with her due to the presence of the two higher-ranking males. However, discord between the two

Fig. 8.11 *Kasonta* chases *Sobongo* in a tree.

Fig. 8.12 *Kasonta* performs a charging display on the ground.

senior males brought him a golden opportunity. When *Kasonta* interfered with his intentions to mate, *Kamemanfu* would rush to *Sobongo*'s side, and *Kasonta* had to fight with the *Sobongo–Kamemanfu* coalition. On the other hand, when *Sobongo* interfered, *Kamemanfu* would rush to *Kasonta*'s side. *Kamemanfu* held the casting vote over the competition. As soon as *Kamemanfu* seemed to be going to *Sobongo*'s side, *Kasonta* desperately tailed *Kamemanfu*, as an infant follows its mother during weaning (Fig. 8.17). After 5 March, *Kasonta* was always following

Fig. 8.13 *Kasonta* and *Kamemanfu* pant-hoot together.

Fig. 8.14 *Kamemanfu* grooms *Sobongo*.

Kamemanfu, and when he noticed that *Kamemanfu* had gone out of sight by moving ahead to search for another food source in the bush, he desperately looked for him, often grimacing or emitting a short scream.

Consequently, nobody interfered with *Kamemanfu*'s mating with *Wasalamba*. *Kamemanfu* gained the highest share of mating among the three adult males, attaining over 50 per cent of all mating (Fig. 8.18).

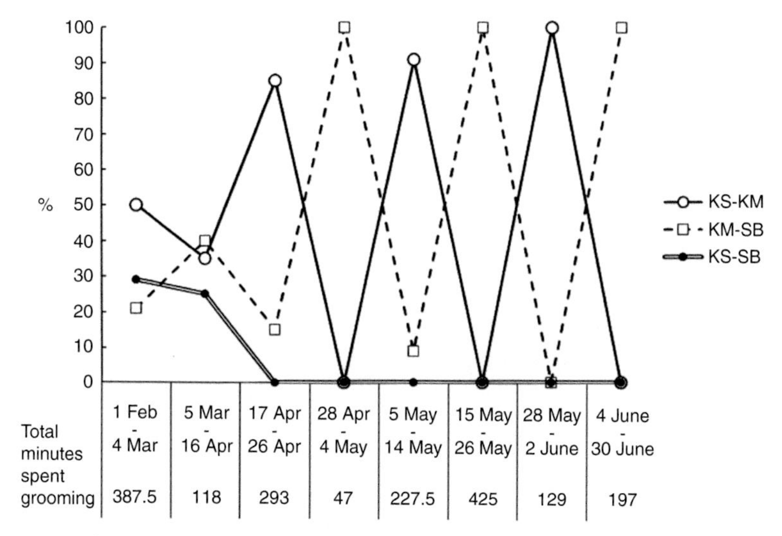

Fig. 8.15 Grooming relationships among three adult males, *Kasonta*, *Sobongo*, and *Kamemanfu* (from Nishida 1983a).

Fig. 8.16 *Kasonta* hugs *Kamemanfu* near *Wasalamba* (in oestrus).

Fig. 8.17 *Kasonta* grimaces and follows *Kamemanfu* as an infant follows its mother, as soon as *Kamemanfu* seems to approach *Sobongo*.

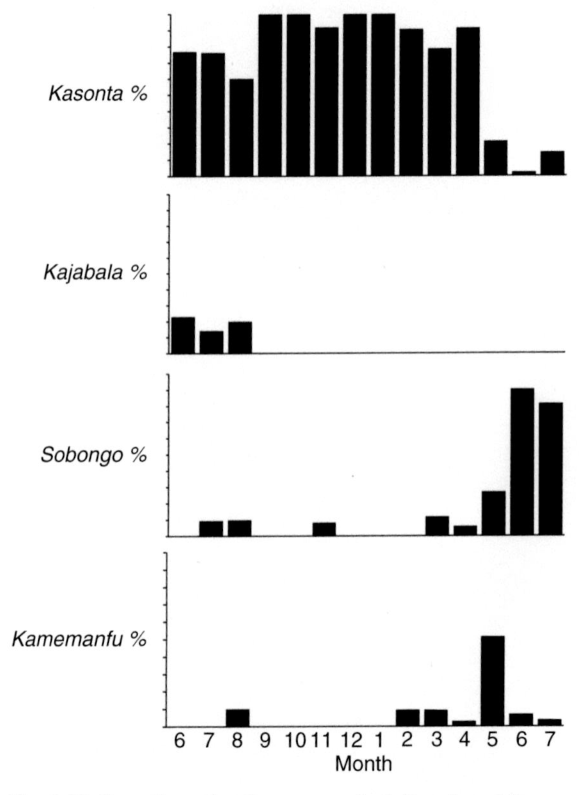

Fig. 8.18 Overall mating frequency of adult males of K-group (from Nishida 1983a).

On 14 May, in the afternoon, fighting and disorder occurred high in the trees along the Mpila valley. After the commotion ceased, *Kamemanfu* and *Sobongo* groomed mutually for 12 minutes, and then *Sobongo* groomed *Kamemanfu* one-sidedly for 9 minutes. Then the two males cooperatively assaulted *Kasonta*. *Kasonta* disappeared from the scene without offering any resistance. He climbed down from the fig tree where the fighting had been raging and fled to the upper stream, without emitting a pant-grunt. By that time, *Kasonta* had received many wounds, in particular heavy wounds on his back and head. *Kasonta*'s flight raised *Kamemanfu*'s status from third to second rank, but ironically robbed him of his previous sexual privileges, returning his share to 5 per cent or so, as it had been before the takeover (Fig. 8.18). With the disappearance of the common rival, *Kamemanfu* ceased to be an essential ally for *Sobongo*. The lost political power led directly to lost sexual privilege.

Kasonta lost out to *Sobongo* and disappeared, and *Sobongo* took his place as alpha male. *Sobongo* was much more powerful than *Kamemanfu*, but whenever they were roaming, *Sobongo*, as the younger of the two, always followed *Kamemanfu*. Accordingly, *Sobongo* became the alpha male, but a leader he was not.

The next year I returned to Mahale for a short follow-up study, after a year of residence in Tokyo. One day in July 1977, I was sitting alone, exhausted after searching in vain for K-group in the higher region of the Mpila valley. I heard an animal approaching me, which I thought was a bushpig or some other large animal that had not seen me. But it was *Kasonta*, travelling alone. He appeared fat and in good health. This was the first time I had ever discovered an ousted alpha male roaming alone on the periphery of the group's range. According to Shigeo Uehara, he was sometimes seen with adult females.

8.3.3 *Kasonta*'s reinstatement and the destiny of K-group males

According to Shigeo Uehara (Uehara *et al.* 1993), at the end of September 1977, *Kasonta* suddenly returned to K-group, and *Kamemanfu* again supported *Kasonta* against *Sobongo*. Although *Sobongo* did not pant-grunt to *Kasonta*, he was alienated by the coalition and began to roam alone. Thus, *Kasonta* and *Sobongo* apparently 'exchanged' their positions without any substantial fighting.

In the middle of January 1978, however, *Kasonta* disappeared from K-group forever, although he seemed to be in good health when last seen. We do not know why he disappeared. However, as Uehara noticed a great commotion from M-group in the upper valley of Myako directly before *Kasonta*'s disappearance, we suspect that he was killed by M-group's chimpanzees. As Myako valley was within the exclusive range of K-group, this speculation was reasonable. In early February 1978, *Sobongo* returned to the group and began to travel with *Kamemanfu*. However, *Sobongo* also disappeared in the middle of May of the next year, 1979.

The riddle was not solved as to why *Sobongo* dropped out of sight, but there is a strong likelihood that he also was slain by the M-group males. Whatever the case may be, as there was no other adult male left except *Kamemanfu*, he became alpha male by default, unlike males who earn the top position as the result of fierce battles.

8.4 *NTOLOGI*, UNPARALLELED LEADER OF M-GROUP

8.4.1 *Ntologi*'s heyday

When we began to identify most of M-group's individuals in the early 1970s, the group's alpha male was probably *Sankaku* (Fig. 8.19), and following him was *Kajugi*. *Sankaku* occupied alpha status for more than three years and *Kajugi* for another three years. *Sankaku* was in the same

Fig. 8.19 A violent alpha male, *Sankaku*.

Plate I.1 All African apes perform the 'aeroplane' pattern.

Plate 1.2 Mahale Mountains viewed from the south.

Plate 1.4 Miombo woodland on the slopes of Mt Pasagulu.

Plate 1.5 Kasoje forest.

Plate 1.6 Mountain forest near the summit of Mt Nkungwe.

Plate 1.7 Mountain grassland decorated with *Protea gaguedi*.

Plate 1.8 Mountain grassland decorated with *Helichrysum*.

Plate 1.9 Solid-stemmed bamboo (*Oxytenanthera abyssinica*).

Plate 2.3 Chimpanzee digging for *Aeschynomene* root.

Plate 2.5 Fruits of *Saba comorensis* (courtesy of Michio Nakamura).

Plate 2.7 Fruit of *Voacanga lutescens*, with the pulp eaten by chimpanzees.

Plate 2.11 Male chimpanzee swallows the leaf of *Commelina*.

Plate 2.13 Red colobus sitting on the ground.

Plate 2.16 Ant fishing.

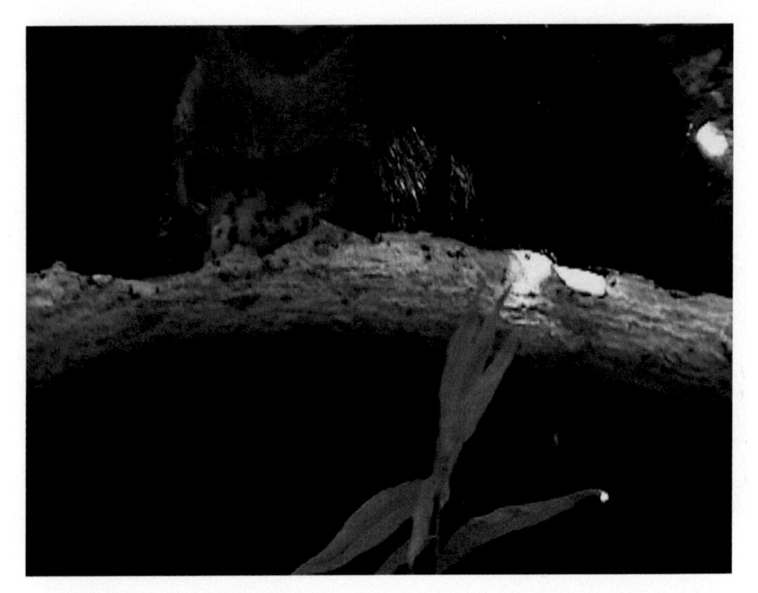

Plate 2.18 Adult male, *Carter*, gnaws a dry, stout stick to eat the eggs of *Crematogaster* ants.

Plate 2.19 Mongo women fishing for termites.

Plate 3.2 Mother, *Chausiku*, transports her newborn son on her head.

Plate 3.13 Weanling, *Michio*, monitors his mother's response to his temper tantrum.

Plate 4.1 Adult male slaps at a fresh leopard carcass.

Plate 4.2 Many chimpanzees gather around an old female, *Fatuma*, who picks out sand fleas.

Plate 4.11 Two adolescent females play-wrestle on the ground.

(a)

Plate 5.6a, b, c Leaf-clip display.

(b)

(c)

Plate 5.6a, b, c (cont).

Plate 6.7 Myako camp, when it was thriving in 1979.

Plate 8.14 *Kamemanfu* grooms *Sobongo*.

Plate 8.24 *Shike* with meat, and *Kalunde*.

Plate 8.26 *Ntologi* ignores *Nsaba*, who is following behind.

Plate 8.28 *Ntologi* unconscious.

Plate 8.29 *Kalunde* puts his teeth on the back of *Dogura*.

Plate 8.31 *Kalunde* lets females eat a colobus carcass (courtesy of Kozo Yoshida).

Plate 8.32 *Ntologi* and a meat-eating cluster (courtesy of ANC Productions).

Fig. 8.20 An old male, *Kalindimya*.

class as *Kasonta*, a tyrant. Something I found curious was that if the old-aged male *Kalindimya* (Fig. 8.20) turned up, *Sankaku* let the older chimpanzee snatch all the sugarcane and bananas that he had kept to himself, just as *Kasonta* allowed his bounty to be snatched by a very old but lower-ranking male, *Kasagula*, in K-group (see above).

Ntologi (Fig. 8.21) was, without a doubt, an alpha male extraordinaire, although he, like *Kasonta*, was afraid of humans. He reached the top position in July 1979. Unfortunately, we did not know how he took over the top position from *Kajugi*, because we researchers were then taking turns doing research in the town of Kigoma. In July, when Kenji Kawanaka and his wife Hatsuko left Mahale, *Kajugi* was the alpha male. When Yukio Takahata, Mariko Hasegawa, and I later arrived to take their place at Kasoje, we found that *Ntologi* had already begun behaving as an alpha male, and *Kajugi* was an exile, travelling alone, usually at the southern periphery of M-group's range. From then on, *Ntologi* maintained the alpha position for an unbroken 12 years. After being dethroned from his alpha status by the two-male coalition of *Kalunde* and *Nsaba* around March 1991, he bounced back to the top status ten months later and maintained power for three more years.

All of the alpha males I came to know, except for *Sobongo* and *Kamemanfu*, were large in body size. *Kasonta* weighed in at 57 kg, and *Ntologi* and *Sankaku*, who did not even fit on the scales, probably weighed close to 60 kg each, in contrast to *Sobongo* and *Kamemanfu*,

Fig. 8.21 *Ntologi* in his prime.

who weighed 47 kg and 35 kg, respectively. Both *Kajugi* and *Kalunde* had large builds. As two of the famous alpha males from Gombe National Park had rather small body types, Jane Goodall concluded that body frame had nothing to do with becoming top male. Recently, Pusey and colleagues (Pusey *et al.* 2005) reported that dominance and body-weight are correlated among adult females but not among adult males at Gombe. Nevertheless, all else being equal, being larger in body size may be advantageous.

8.4.2 *Ntologi*'s politicking tactics

Ntologi was popular in the group and had many hangers-on around him (Fig. 8.22). Three senior males, *Kalindimya*, *Rashidi*, and *Kagimimi*, and

Fig. 8.22 *Ntologi* leads a large party in 1989.

three age-mates, *Bakali*, *Musa*, and *Lubulungu*, approached and pant-grunted to *Ntologi* in sequence and then groomed one another. *Ntologi* always enjoyed being at the centre of a big grooming cluster. In particular, *Lubulungu* was *Ntologi*'s best friend. Ambitious but young, high-ranking males were so much overwhelmed by the senior male club that they would not even approach their grooming cluster.

Ntologi had great stamina for intimidation displays lasting as long as several minutes at a stretch, while other males usually stopped after 10–20 seconds. He pulled at a woody vine, threw dry branches one after another, repeatedly rolled rocks, swiftly climbed a tree, and displayed showy brachiation. Moreover, *Ntologi* was strong enough to lift up heavy rocks and throw them into a stream with both hands. The high splashing water and sound were enough to intimidate other chimpanzees looking on.

Ntologi was also very skilful in drumming displays. In the camp, as described already (Section 8.1), he not only slapped the wall of the metal house with his hands but also kicked it with his feet. Moreover, he was the only male to visit and drum on two metal houses in each display bout. Of course, staging intimidation displays is a much less risky, though perhaps no less energetically expensive, way to maintain alpha status than directly attacking rivals.

Furthermore, *Ntologi* was not only one of the best hunters but also could snatch a carcass from other chimpanzees. During the late

1980s, more than 30 per cent of carcasses hunted by M-group's chimpanzees went to *Ntologi* (Nishida *et al.* 1992). This is an extraordinary figure, given that the apes sometimes killed up to 11 monkeys in one hunting event. Taking advantage of these resources at his disposal, he shared meat with senior males such as *Kalindimya* and *Kagimimi* and age-mates such as *Lubulungu*, in addition to parous females who were his previous sex partners, including the alpha female *Wakampo*, oestrous females, and his mother. He never shared meat with his rivals, second-ranking male *Bakali*, or third-ranking *Kalunde*, who succeeded to the second rank after *Bakali* relinquished the position due to old age. As soon as *Bakali* descended in rank and was no longer a rival, he was allowed to join the 'senior male club' and *Ntologi*'s meat-eating banquets (Nishida *et al.* 1992). Consequently, in later years, whenever *Ntologi* got a big kill, his male age-mates *Bakali* and *Musa* (*Lubulungu* had died by then) clung to his side, feasting to their hearts' content. *Ntologi* even left the main treats of the kill for them to eat. When they got leftovers, they combined their strength to keep the prize from hungry adversaries. As *Ntologi* could take back the kill from them if he pleased, he lay on the ground without a care in the world while nibbling on scraps. *Bakali* and *Musa* played the role of watchdogs over the meat.

Kalunde challenged *Ntologi* once in 1989, but he was bitten severely around the upper lip by *Ntologi*. After that incident, he did not challenge *Ntologi* but instead seemed content with maintaining the second rank for two years.

8.5 *KALUNDE*'S RULE

8.5.1 *Kalunde*'s takeover and his weakness

Vying for supreme power, male chimpanzees take the stage in an intense drama. *Ntologi* sat on the throne of M-group for just under 12 years. *Kalunde* ascended to the second-rank sometime in 1987.

Miya Hamai (1992) and her research assistant, Moshi Bunengwa, noticed for the first time a strange attitude in *Ntologi* in February 1991: *Ntologi* seemed to make every effort to follow the fourth-ranking *Nsaba* and often successfully associated with him. *Ntologi* even searched for *Nsaba* while showing grimaces, if he noticed the simultaneous disappearance of *Nsaba* and *Kalunde*. It was as if he were afraid of their forming a coalition. Finally, in March 1991, *Ntologi* was ousted from power by this coalition.

Kenji Kawanaka (1990, 2002) studied adult male relationships in 1986 and 1990, when *Ntologi* was alpha and *Kalunde* was beta. In 1986, *Ntologi* and *Kalunde* had peaceful relations. However, in 1990 they stayed together or groomed each other much less often compared with 1986; *Ntologi* displayed against *Kalunde* 14 times more often than in 1986. *Kalunde* pant-grunted to *Ntologi* seven times more often than in 1986. Obviously, *Ntologi* was becoming nervous of *Kalunde*'s strength, and he tried to drive him away in 1990.

When we visited Mahale in August 1991, *Ntologi* roamed alone, keeping a safe distance from the rest of M-group (Fig. 8.23). Still, there were times when he was seen with his age-mate, *Bakali*. Judging from this socialising, he was not completely alienated from the 'old boy's club'.

Strangely, *Kalunde*'s rival as the new leader was not second-ranking *Shike*. *Shike* was not *Kalunde*'s coalition partner, but he was afraid of him, never disobeying him. *Kalunde* did have an insecurity with which to contend, because he was bedevilled by *Ntologi*'s shadow and by two main rivals: the now third-ranked *Nsaba* and the fifth-ranked *Jilba*. The 15-year-old *Jilba* greeted second-ranking *Shike* but never paid respect to *Kalunde*; in fact, he mounted a campaign to belittle the alpha, constantly harassing him but then running away when he elicited a serious response. *Nsaba* sometimes paid respect, but whenever *Kalunde* had meat or kept an oestrous female for himself, *Nsaba*

Fig. 8.23 *Ntologi* travels alone in 1991.

threatened him. *Nsaba* and *Kalunde* had united forces against *Ntologi*, but after the common rival was ousted, the rivalry between *Nsaba* and *Kalunde* came to the forefront. Yesterday's friend became today's foe.

Kalunde was just past his prime at age 28 years, but very well built. The source of his downfall was his lack of self-confidence. This was obvious to me because his display was quick and of short duration, consisting of running, dragging branches, slapping the metal wall with both hands at amazing speed, and quickly throwing small rather than large stones one after another. During rest time, he occasionally tore off pieces of dead bark purposelessly for a long time, with trembling lips. This revealed his nervousness or over-sensitivity, which seemed to have been brought on by gaining the top position in the male hierarchy. When it came to confrontation, he exposed his gums, grimaced, and then rushed to a nearby adult male or even female, mounted, placed his teeth on the partner's back, and thrust against him or her. 'Mounting' serves the purpose of reclaiming one's self-confidence and soliciting the mounted individual's backing. Apparently, he could not keep a stiff upper lip. *Kalunde* mounted others so often that one of the video crew hollered boldly, 'Oh look, *Kalunde*'s packing fudge again.' This expression coming from a young woman always sent me into fits of laughter. Kalunde's frequency of mounting was highest during and just after his period of supremacy.

There was a bizarre dead-locked triangle going on among the three top-ranking males. *Kalunde* was overwhelmingly violent toward second-ranking *Shike*, and when *Kalunde* charged, *Shike* stood petrified after pant-grunting. On the other hand, *Shike* was undeniably harsh to *Nsaba*, and if *Shike* charged at him, *Nsaba* not only pant-grunted but also let out shrieks, running this way and that seeking refuge.

However, *Nsaba* was not at all afraid of *Kalunde*. *Nsaba* even showed charging displays against the alpha male. When *Nsaba* began performing an intimidation display directed at *Kalunde* or without an obvious target, *Shike* appeared to consider this to be a challenge to his status and flew at *Nsaba*, chasing him away. Consequently, *Kalunde* maintained top status. It was not clear whether or not *Shike* was actually *Kalunde*'s ally, but since *Shike* was extremely harsh to *Nsaba*, it appeared as though he at least backed *Kalunde*. Therefore, *Kalunde* did not dare to touch or even to glance at a colobus carcass in the possession of *Shike* (Fig. 8.24), although it would be easy for him to do so. This was probably because *Kalunde* was afraid to give *Shike* cause for animosity by plundering his goods.

Fig. 8.24 *Shike* with meat, and *Kalunde*.

Ntologi's old friend, *Musa*, was now an ally of *Kalunde*. *Musa* did not fight *Kalunde*'s battles for him, but at the very least he took *Kalunde*'s side in fights. Before *Kalunde* began to fight, he grabbed *Musa* around the hips and mounted him from behind, relieving a little anxiety.

8.5.2 Gang attack on *Jilba*

On 7 October 1991, when *Kalunde* was alpha and *Ntologi* was still wandering alone, a rare incident occurred. Five adult males, two adult females, and an adolescent male beat-up the 16-year-old male, *Jilba*. A synopsis of the incident follows (Nishida *et al.* 1995).

That morning there was a hunting expedition, in which a female, *Opal*, caught a colobus monkey. The alpha male, *Kalunde*, robbed her of her catch, and a meat-eating cluster centring around *Kalunde* began in the treetops, with *Kalunde* sharing the meat with five or six other chimpanzees.

About 11:15 a.m. *Jilba* showed up, approached close to *Kalunde*, and twice performed a charging display in the canopy. Two minutes later, *Kalunde* alone chased after *Jilba*, both fell to the ground, and *Kalunde* caught *Jilba* in some bushes. The chimpanzees from the meat-eating cluster followed after, and those who first heard the screams rushed *en masse* to the scene.

Kalunde had *Jilba*'s lower body pinned, and every member of the gang, the second-ranking male, *Shike*, the fourth-ranking male, *Aji*, the young adult male, *Bembe*, the old male, *Musa*, and two adult females,

Wakusi and *Happy* (but excluding the adolescent male *Hamby*, *Happy*'s son), took turns tugging and nipping at *Jilba*'s back, arms, and legs. *Jilba* had no way to escape, so he stood on all fours, helplessly being bitten, and then defecated.

Other chimpanzees looked down on the scene from the treetops, giving 'waa' barks that could have been positive or negative, the majority of them sounding fearful. About two minutes after the commotion, *Jilba* escaped. But the fear calls went on for 14 minutes. The only two adult males at the scene of the violence who did not partake in beating *Jilba* were third-ranking *Nsaba* and a young adult male, *Toshibo*. *Jilba* went missing for 50 days after his brush with death. Never again have I seen such a mob attack!

Up to that point, no field researcher had reported such a case, in which a gang carried out a beating of a male of the same group, other than the former alpha male (see below). Why did *Jilba* meet with such a brutal attack? Could it have been ostracism?

For a few months before the incident, *Jilba* displayed peculiar behaviour. He was not paying respect to anyone except the second-ranked *Shike*. By paying respect to *Shike*, I mean that I documented *Jilba*'s pant-grunting at *Shike* only three times in three months, and these cases resulted only from *Shike* threatening *Jilba*. On the other hand, *Jilba* was not pant-grunted to by males considered lower in rank than him.

A noteworthy point is that *Jilba* threatened and attacked the alpha male, *Kalunde*, many times. When *Jilba* squared off with *Kalunde*, it was either his undoing or he successfully carried out his fight. *Kalunde* then grabbed the haunches of *Aji*, *Musa*, or some other male and mounted and put his teeth on their backs, in order to regain his self-confidence.

Kalunde often groomed other males in the group but was rarely seen to groom *Jilba*. However, when he did groom with *Kalunde*, *Jilba* always behaved in a strange way; that is, when *Kalunde* groomed him (Fig. 8.25), *Jilba* stared at the ground and groomed a dead leaf! Moreover, there were times when young adult males (especially *Jilba* and *Shike*) assaulted adult females without justification.

The background at the time cannot be overlooked, for, when *Kalunde*'s status was unstable, contention and menace among the males became an everyday occurrence. Perhaps amplified agitation at meat-feeding times derived from this seed of tension, as it appeared the group sought to vent their dissatisfaction, provoked by *Kalunde*'s hostility.

Fifty days after *Jilba* had suffered the brutal gang beating, he and the former alpha male, *Ntologi*, who had been driven off, were sighted

Fig. 8.25 *Kalunde* grooms *Jilba*, who is ground-grooming.

together. At the start of 1992, when *Ntologi* made his comeback in M-group, *Jilba* joined him. It appeared that both of them sought to return to M-group, and thus they availed themselves of each other's support. An adult male cannot live alone, because he will find only enemies once he ventures outside the natal group's territory.

Several cases of gang attack among wild and captive chimpanzees have been reported. One occurred at Gombe, when a patrol squad consisting of male members from the Kasakela group wandered into a neighbouring Kahama group range, assaulting and then slaying its members, one after another; it looked as though a war had broken out. The Kahama group had originally split off from the Kasakela group, and it then consisted of eight adult males and one middle-aged female (Goodall *et al.* 1979).

Another case of gang attack occurred at Gombe, involving a former alpha male who had been ousted from the group and a squad based around the current alpha male. The alpha male, *Goblin*, lost his alpha status when the second-ranked male launched an attack, mauling *Goblin*'s testicles so severely that he sustained near-fatal injuries and had to undergo medical treatment from a veterinarian. Upon recovering and returning to the group, he fell victim to a group assault carried out by six adult males, including the new alpha male (Goodall 1992).

In captivity, two adult males in coalition killed another adult male and removed his testicles (de Waal 1998). More recently, an adult male was killed and his testicles were found on the ground at Kibale forest (Muller 2002). In ancient China, one of the punishments for a political criminal was to remove his testicles. Kano also noticed that some of the males in his study group of bonobos at Wamba lacked testicles (Kano 1984). It appears that the same idea occurs to male humans, chimpanzees, and bonobos, although the reason for the lack of testicles among Wamba's bonobos is unknown.

A month prior to the gang attack against *Jilba*, *Kalunde* and his crew (three males and at least two females) ganged up on the dethroned *Ntologi*, but he made a narrow escape. Had he been caught, it would have been another case of a gang beating. I was surprised to see all the males and females that had pant-grunted earnestly to *Ntologi* when he was alpha now openly attacking him, supporting the new alpha, *Kalunde*.

It is difficult to determine whether the incidents we witnessed were behaviours meting out a temporary rebuke, such as that given to a youth who is ill-mannered and lacks respect for the most basic forms of social etiquette, or a permanent punishment to the order of ostracism or banishment, similar to the case of the former alpha male.

I support the ostracism theory. Would not *Jilba*'s return to M-group have been out of the question, had not the unusual circumstances of *Ntologi*'s comeback been an influence?

According to Shigeo Uehara, the loathed *Jilba* was the repeat victim of another gang beating in June of the following year, 1992. Again, it was the work of *Kalunde* and his cronies; *Jilba* suffered lacerations on his anus and the middle finger of his left hand.

8.5.3 What *Jilba*'s case teaches us

These rare happenings lend credence to the notion that humans and chimpanzees share several common psychological traits.

First, chimpanzees make a clear distinction between members and non-members of the group. The apes not only fear and loathe other groups, but they seem to have a desire to eradicate them completely. When one group becomes aware of members of another group nearby, their hair stands on end, they display looks of horror, they cling to their cohorts, and they experience bouts of diarrhoea. Members of the same group are occasionally treated as members of an outside group, namely 'ostracised' individuals such as former alpha males, or *Jilba*.

Second, the alpha male has great influence over *who* is regarded as a group member and who is not.

Third, there is a tendency to follow the crowd, and, in particular, a strong tendency for the crowd to take the alpha male's side. This reflects a policy of supporting the stronger individual, or 'winner support'.

Fourth, 'you scratch my back and I'll scratch yours' (*quid pro quo*) behaviour prevails, being seen in grooming sessions or when a female permits a male to copulate with her, with the mutual understanding that she will be entitled to a share of meat or to other benefits. Fifth, treachery is avenged by assault, with a tendency to follow the 'eye for an eye, tooth for a tooth' principle (de Waal 1989). Such behavioural characteristics as revenge or retribution are demonstrated in a gang attack.

8.6 *NTOLOGI* RESTORED

8.6.1 A change of regime

At the end of 1991, *Shike* fell victim to a serious illness and his status as second-in-rank was lost as he plummeted to the bottom of the male dominance hierarchy. According to Kazuhiko Hosaka (Hosaka 1995b; Hosaka & Nishida 2002), with this shift, *Nsaba*'s status was fortified and no one could tell who was now the alpha male, *Kalunde* or *Nsaba*. *Nsaba* no longer paid respect to *Kalunde*, and neither pant-grunted to the other. Eventually, nothing was heard of *Shike*, and the word around the campfire was that he had died.

Exactly at this moment, to everyone's amazement, *Ntologi* came back! Hosaka, who was then supposed to return to Japan, telephoned, asking me to allow him to stay longer to take advantage of this fascinating opportunity. I immediately responded: 'Yes, you stay and observe as much as you like.'

First, *Ntologi* made amends with *Jilba* and with his age-mate *Musa*, securing a spot in the 'old boy's club'. *Kalunde* and *Nsaba* built an alliance against *Ntologi*, but it did not last long. As *Nsaba*'s relative status rose, even if *Kalunde* were to win out over *Ntologi*, he would have to be content with answering to *Nsaba*. On the contrary, if *Kalunde* sided with *Ntologi*, he would come out ahead as victor over *Nsaba*.

Kalunde applied some contradictory tactics. One day *Kalunde* sided with *Nsaba* and rivalled *Ntologi*, but the next day *Kalunde* partnered up with *Ntologi* to cooperatively launch an attack on *Nsaba*. This was the reappearance of the 'alliance fickleness strategy' seen earlier in K-group. Ultimately, *Kalunde* joined *Ntologi* in employing the 'winner support' or 'stronger support' strategy that males commonly engage in, and thus March 1992 saw the resurgence of *Ntologi*'s regime.

Kalunde never did succeed in turning the poison named *Ntologi* into medicine for himself, since *Kalunde* lost out even to *Nsaba*, sliding down to third in rank.

What mystifies me is how *Ntologi* had the foresight to time his resurgence as he did. Since he could no longer hear *Shike*'s unique intimidation displays and pant-hoot vocalisations, he may have assumed that *Shike* had disappeared. But even with *Shike's* disappearance, would *Ntologi* also have been able to deduce the mounting tension between *Kalunde* and *Nsaba*? *Ntologi* was 10 years older than *Kalunde* and close to 20 years older than *Nsaba*. Without the splitting of *Kalunde* and *Nsaba*, *Ntologi*'s resurgence would have been impossible. It does not seem realistic to think that *Ntologi*'s return was pure coincidence, but on the other hand, saying that it was a calculated strategy would be reading too much into the events.

However, it may be that *Ntologi* obtained the necessary information for his return to power. After *Nsaba* and *Kalunde* separated, *Ntologi* must have heard more screams from *Kalunde* and must have noticed the decrease in communal pant-hooting between *Kalunde* and *Nsaba*, which is the signal of a firm coalition. When we were following *Ntologi* in his lonely days of 1991, he was usually not far from M-group's core party, including *Kalunde*; he kept several hundred metres from them and continued to listen to their calls. For example, one day in August 1991, when the core party was staying in the Kasiha valley, *Ntologi* remained sitting at the top of a nearby hill, listening to calls from the party, changing sitting positions and listening directions. That is, he seemed to be collecting information on the social dynamics from a distance (Hosaka 1995b; Hosaka & Nishida 2002). In addition, *Ntologi* may have gained more confidence when he succeeded in reforming the alliance with *Jilba* and *Musa* (Hosaka 1995b; Hosaka & Nishida 2002).

8.6.2 The ingenuity of *Ntologi*'s rule

Upon my return to Mahale in August 1992, *Ntologi* was back on the throne. His wonders never ceased to amaze me. *Kalunde* had again sided with *Nsaba* in making vain attempts to threaten *Ntologi* and then made an about-face in launching an attack on *Nsaba* with *Ntologi*'s aid. But in the end *Kalunde*'s activity was nothing more than a storm in a teacup, as *Nsaba* had lost the drive to stage any more attacks on *Ntologi*.

Ntologi applied appeasement tactics to *Kalunde*, but with *Nsaba* he launched an all-out 'intimidate and ignore' campaign. When he met *Kalunde*, he initiated long hours of grooming, cleaning *Kalunde*'s face with an obsessive passion. He did not stop at removing hardened bits of fruit juice from *Kalunde*'s chin with his finger, lips, and incisors, but went all-out, picking *Kalunde*'s nose and then eating what he removed. Because of this splendid service, *Ntologi* was always picking up *Kalunde*'s cold. At the same time, *Ntologi* kept a watchful eye on *Nsaba*, terrorising him with displays and so on. Even when *Nsaba* came to groom *Ntologi* as a peace offering, *Ntologi* totally ignored or 'snubbed' him (Fig. 8.26).

These tactics of *Ntologi*'s were labelled 'separating intervention' by Frans de Waal (de Waal 1982). He would apparently charge into the grooming session between any pair of adult males in order to interfere with it. He seemed to loathe intimate relationships among other males and interfered jealously. However, the foci of *Ntologi*'s intervention were second-ranking *Nsaba* and third-ranking *Kalunde* (Fig. 8.27), as they potentially were the most dangerous coalition against him (Nishida & Hosaka 1996).

One day when I was tracking *Ntologi* as my target, a large party from M-group was heading north on the path we called 'Route 1'. He had just outrun three adult males and was plodding up the path on his

Fig. 8.26 *Ntologi* ignores *Nsaba*, who is following behind.

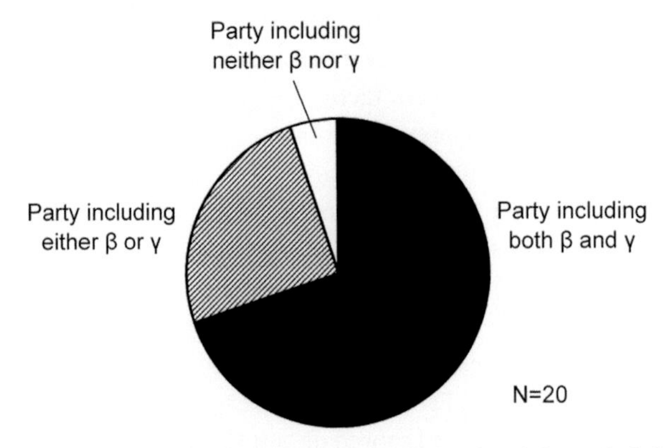

Fig. 8.27 Intervention of groom formation by alpha male (after Nishida & Hosaka 1996).

own about 50 m from the group. Then, he sat down in the brush, a little off the path. He kept looking back down the path; it seemed as though he was waiting for the other males to catch up with him, for even as the females passed by continually, he would not budge. After about ten minutes, *Ntologi*'s hair stood on end and he headed southwards, back down the path. Just as I had thought, and *Ntologi* had thought too, those three males had started a grooming party. *Ntologi* rushed at the group as expected, and chased them into the bushes.

This account may seem a bit over-personified, but I maintain my interpretation of the events with conviction. From the moment *Ntologi* sat on that path glancing back, I predicted that those three males were holding a grooming party and that this activity would compel *Ntologi* to turn back and rush at them, in order to disperse the gathering.

8.7 *NTOLOGI* LOSES AGAIN AND *NSABA*'S HEGEMONY BEGINS

8.7.1 *Ntologi*'s pant-grunting

According to Linda Turner (1995), *Ntologi*'s second tenure as alpha male was cut short by the attack from *Nsaba* on 14 April 1995. Linda's summary of the account from Rashidi Kitopeni and Moshi Bunengwa reads:

> *Ntologi* and *Nsaba* fought and fell through the branches of a tree. *Ntologi* was forced to the ground by *Nsaba*, who showed no fear as *Ntologi* climbed back into the tree. *Ntologi* was again forced to the ground and appeared to

> be seriously injured. His left eye was badly lacerated and he seemed to be in pain, repeatedly covering the eye with one hand and holding his head. He climbed into a tree and joined two of his supporters, the young adult males *Toshibo* and *Aji*. One of them had managed to capture a colobus monkey, and they seemed to pay little attention to *Ntologi* as they concentrated on eating the meat. *Nsaba* lay on the ground a few metres from the tree in which the three males sat. When he suddenly rose to his feet and directly approached the tree, *Ntologi* quickly descended and hurried to meet him with vigorous pant-grunting.

In the afternoon of 19 April, *Ntologi* and *Nsaba* were seen to fight again, and *Ntologi* disappeared and was not seen until 18 May; his status in the group was unclear. As a consequence, *Nsaba* became alpha. The new wave of young adult males, such as *Toshibo*, *Aji*, *Jilba*, and *Bembe*, had grown up. *Kalunde* apparently kept a high rank in the dominance hierarchy by maintaining a strong alliance with *Nsaba*. However, on a one-on-one basis he appeared to be subordinate to most of the young adult males, because it was *Kalunde* who screamed and retreated during display contests. Although *Kalunde* screamed loudly, he never pant-grunted to these young adult males.

I will never forget the last field season when I was able to watch *Ntologi*. When I arrived at Kasoje on 20 September 1995, *Ntologi* had already returned to M-group. His return was on 24 August 1995, but this time he returned as a ruined exile. He pant-grunted to every adult male he encountered. This was the first time I witnessed him pant-grunting, because I had observed K-group most of the time before 1979, when he became alpha. It was an unbelievable scene to see him run away screaming and climb a tree whenever a young adult male charged at him. In meat-eating events, the only thing he could do was to beg *Kalunde* for meat, rob a tiny piece from an adult female's mouth, or recover scattered pieces one by one from the ground, behaviour typically performed by immature chimpanzees! I precisely judged *Ntologi*'s status as the lowest-ranking adult male but higher than any other chimpanzee, because he still chased away *Dogura*, the oldest adolescent male, and forced him to pant-grunt.

Although *Ntologi* lost his high social status, he was nevertheless popular among M-group members. He had no difficulty in obtaining grooming partners. Moreover, he seemed to maintain some political power. On 21 September 1995, when *Ntologi* was in a grooming cluster with high-ranking young adult males such as *Toshibo*, *Jilba*, and *Bembe*, the coalition of *Nsaba* and *Kalunde* charged at them. This was a typical communal separating intervention against *Ntologi* and the young adult males.

A few days later, I observed an interesting development in male coalitions. First, probably in anticipation of support from *Nsaba*, *Kalunde* charged the second-ranking *Toshibo*. In response, *Toshibo* charged *Nsaba*! Seeing this, *Kalunde* and *Ntologi* supported *Toshibo* against *Nsaba*! *Kalunde* betrayed *Nsaba*! Faced with the communal attack by three adult males, the isolated *Nsaba* began to retreat while screaming loudly. Meanwhile, *Aji*, *Nsaba*'s peer and friend, began to scream. *Toshibo* bit *Aji* on his waist. After some more commotion, *Toshibo* and *Nsaba* groomed each other and reconciled.

On 28 September, *Toshibo* and *Jilba* met on an observation path, and were then joined by *Nsaba* and *Kalunde*. A display contest began between *Nsaba* and *Toshibo*, and *Kalunde* displayed on the side of *Toshibo*. *Kalunde* then moved to the side of *Nsaba* and seemed to change the target from *Nsaba* to *Toshibo*. At this moment, *Ntologi* emerged and chased *Nsaba* while grimacing and screaming loudly. *Ntologi* frequently looked back at *Toshibo* while chasing *Nsaba*, and then all the rest (*Toshibo*, *Kalunde*, and *Jilba*) followed suit and supported *Ntologi* against *Nsaba*. Poor *Nsaba* was totally alone and seemed to be at a loss about what to do. These episodes give eloquent proof of how *Ntologi* was still able to wield political power, or the power to mobilise other adult males, even after he became the lowest-ranking adult male.

8.7.2 The death of *Ntologi*

On 14 November 1995, Noriko Itoh and Hamisi Bunengwa were looking for chimpanzees in a higher part of M-group's range. When they had almost given up the search, they heard faint calls from a distance. Unexpectedly, calls came from the foot of the mountains, around Route 1. When they arrived at the lower Kasihamto valley, they found *Ntologi* lying on the ground on his left side, with his eyes closed (Nishida 1996a). Present were a young adult male, *Aji*, four adult females, an adolescent female, two juvenile males, and a juvenile female. Most were in a tree, while a few were on the ground. *Aji* leapt down to the ground and slapped *Ntologi*'s back before grooming him. At that time I was in the camp, and when Hamisi brought me the news, I rushed to the scene.

Ntologi was breathing but in a coma and so did not move (Fig. 8.28). He had many wounds on his body, right and left thumbs and third digit of the left hand, a removed nail on the fourth digit, two lacerations on the left shoulder, an injury on the left side of the belly and top of the right thigh, five teeth marks on the right side of the back,

Fig. 8.28 *Ntologi* in a coma.

and three teeth-marks on the left side of the chest. His left and right ears and his lips were also injured. *Aji* was performing charging displays and almost lunged at *Ntologi* when I intervened. Three chimpanzees peered at *Ntologi* from 5 m above the ground. While I was investigating *Ntologi*'s body, *Aji* tried to make charging displays against him three times, which I interrupted.

We had no idea how to treat him, but we thought that if we left him in the bush, he would be devoured by leopards, bushpigs, or army ants. We prepared a stretcher and brought him to our camp. When we began to carry him, adult females and adolescent chimpanzees began to emit 'fear calls' and *Aji* resumed charging displays against us.

We kept *Ntologi* in one of the rooms of our KUAPE (Kyoto University African Primate Expedition) house. Since *Ntologi* was in a coma, he could not eat anything. Therefore, it would have been in vain, even if we had prepared a banana containing an antibiotic. He occasionally gave a loud groan that resembled a human call. He breathed 26–32 times per minute, which was a greater than usual frequency (19 times per minute) for chimpanzees (Flindt 2006). We sought the help of a veterinarian from the headquarters of the national parks at Arusha through the call station of Mahale Park, but no one could come.

At 03:23 hours the next morning, *Ntologi* died. No one knows how he was attacked. However, from the many wounds received and given that the scene of the crime was in the central part of M-group's territory, I speculate that he was subjected to a gang attack by his groupmates, such as *Nsaba*. None of the nine chimpanzees found at the site of violence, including *Aji*, had any wound, and thus I believe they arrived after *Ntologi* was attacked. This incident reminded me of how the Cold War president of Romania was killed by people who had resented his selfish and violent rule. Likewise, *Ntologi* had reigned selfishly over the group for 15 years, and thus other members may have held grudges against him.

8.7.3 The disappearance of adult males

Ntologi was fatally attacked on 14 November 1995. In December we celebrated Mahale's thirtieth anniversary by holding an international symposium in the capital city of Dar es Salaam. Sponsored by the Japan International Cooperation Agency, we set up an exhibition of photographs and explanatory panels at the National Museum, while we held a two-day discussion on the ecology and conservation of animals – chimpanzees in particular. In October, we also had an in-house celebration at Kasoje with our field assistants and a filming team.

After the death of *Ntologi*, strange incidents occurred one after another. First, four young adult males, *Bembe*, *Toshibo*, *Aji*, and *Jilba*, and an adult female, *Wakilufya*, disappeared in succession between the mid-December 1995 and early January 1996. The alpha female *Wakampompo*, her juvenile daughter and infant son, and two adolescent males, *Nick* and *Iwan*, also disappeared by the end of 1996. Most curiously, the alpha male, *Nsaba*, disappeared on 31 December 1996. Only three corpses were recovered; one was *Wakilufya*, and another was probably *Jilba*, given the lack of canine teeth in the corpse. I was in Japan when I heard this sad news. We lost as many as 15 chimpanzees within about 13 months. I imagined that the group might have fissioned, and in 1997 Noriko Itoh and I investigated the southern periphery of M-group's territory, but we found no splinter group. Therefore, we could only speculate that something like an epidemic had afflicted M-group. However, I was not able to visit Mahale during the whole of 1996, and we had no other means to investigate the reason for the mass disappearance.

8.7.4 *Kalunde*'s second rule

The loss of the alpha male, *Nsaba*, and other high-ranking males brought the 33-year-old beta male, *Kalunde*, up again to the alpha position. This time *Kalunde* seemed to be more confident than in 1991, perhaps because all of the young high-ranking adult males had suddenly disappeared. The adult male next oldest to *Kalunde* was about 14 years younger – the 19-year-old *Masudi*. No one predicted that *Masudi* would become high-ranking because he was far from muscular. I imagined that *Kalunde* would enjoy top status for at least several years. However, this did not turn out to be the case. Eighteen-year-old *Fanana* soon found a good chance to usurp the alpha position.

Kalunde spent two years as the alpha male, and during this period the 27-year-old female *Nkombo* almost married him. We observed *Kalunde* and *Nkombo* travelling together most of the time. This was the first case of the formation of a strong pair-bond between a particular adult male and a particular nulliparous adult female that we had ever seen. Although both tended to follow each other, it was *Nkombo* who was mostly responsible for maintaining the long-term association. When I investigated who followed who from 1999 to 2004, *Nkombo* followed *Kalunde* on 70 per cent of 29 occasions.

8.8 THE AGE OF NEW LEADERS

8.8.1 *Fanana*'s power takeover and female intervention

Fanana challenged *Kalunde* for alpha status in November 1997. He scarcely ever enlisted support from the third-ranking *Dogura*, who was 16 years old, but instead ventured to attack all comers by himself. It was *Kalunde* that enlisted the support of *Dogura*; when *Fanana* charged at *Kalunde*, he ran to *Dogura*, hugged and mounted him from behind, and repeatedly thrust against him, while putting his teeth onto his back (Fig. 8.29). This was his typical support-enlisting behavioural pattern. However, *Dogura* seemed uninterested in supporting either party. He kept a neutral position between *Fanana* and *Kalunde*.

In response to *Fanana*'s charging displays (Fig. 8.30), *Kalunde* also counter-attacked and performed charging displays by running bipedally and throwing rocks. A startling phenomenon was the females' intervention in the fighting between the two highest-ranking males. Three adult females, *Pinky*, *Gwekulo*, and *Nkombo*, joined *Kalunde*'s side, and all of them barked at *Fanana* when he charged at *Kalunde*. A few times they even chased after him, although none of them did more than give vocal support most of the time. Due to these females' intervention, it took three days for *Kalunde* to finally surrender to *Fanana*.

Fig. 8.29 *Kalunde* puts his teeth on the back of *Dogura*.

Fig. 8.30 *Fanana* performs a charging display.

This is the only case in which we have witnessed physical intervention by females in the top males' confrontations. I do not recollect precisely when the strong male–female bond between *Kalunde* and *Nkombo* was formed, but at least by 1996 their bond had become very strong, as mentioned above. *Kalunde* was also popular among other adult females. He was generous in sharing meat with adult females; he sometimes ate only a little while keeping a colobus carcass mostly intact to deliver to

Fig. 8.31 *Kalunde* lets females eat a colobus carcass (courtesy of Kozo Yoshida).

adult females (Fig. 8.31). This shows that being possessor of a carcass was important, because it granted him the role of meat provider.

When *Sobongo* attacked *Kasonta* in K-group, *Sobongo*'s mother helped only by barking, which was ineffective. This was the only case of intervention by a mother, among the six times that a mother was alive during a power takeover involving her son.

For chimpanzees in captivity, de Waal reported cases of interference by adult females and even stated ideas such as 'democracy' being maintained by the females' intervention (de Waal 1982). In the wild, on eight of ten occasions of power takeover, females did not intervene. The discrepancy was likely brought out by the differences in the environments where the chimpanzees were living. There are good reasons why females typically do not intervene in conflicts of power takeover among adult males. First, females are weaker than males, and thus intervention would not be effective. Moreover, female intervention risks the welfare of infants. This situation differs from that of bonobos, who have smaller sex differences in body size: bonobo mothers support their sons in their efforts to acquire the alpha position.

8.8.2 *Fanana*'s control and *Kalunde*'s cooperation

We do not know *Fanana*'s birthdate, because he suddenly appeared in 1988 as an adolescent male (about ten years old). Probably, his supposed mother *Fahari* transferred earlier into M-group with her offspring, but she was so shy that she probably roamed in the southernmost part of M-group's range, which prevented us from identifying her and her family earlier than 1988. Such male transfer has been confirmed four times so far: *Masudi* at the age of five years from K-group (Takahata & Takahata 1989); *Dogura* at the presumed age of seven; *Hit* at the presumed age of eight; and *Fanana*. On the other hand, *Chausiku* of K-group transferred in 1979 to M-group without being accompanied by her five-year-old weanling son, *Katabi* (Hiraiwa-Hasegawa & Hasegawa 1988). It was remarkable that *Katabi* apparently refused to accompany his mother in her transfer to the neighbouring M-group. Transfer of males seems to be permitted by resident adult males only if the immigrants are juveniles and are accompanied by cycling mothers who resume oestrus. Therefore, although chimpanzees are typically patrilocal, it is unlikely that only one type of Y chromosome is maintained in a single unit-group.

Fanana's promotion to alpha was something we had not predicted, because his charging display was not so impressive: he usually slapped the ground only with the left hand repeatedly and then kicked the buttress of a tree. At the camp, he stood bipedal and slapped the wall of the metal house with only the left hand slowly, and he rarely kicked it. His display was far from impressive.

However, after he took over alpha status, his reign made our eyes widen. He began to throw large rocks into streams, with both hands in a bipedal posture, just like *Ntologi*, *Kalunde*, and *Nsaba*.

A defeated alpha male does not pant-grunt to the victor, but runs away to the periphery of the group's range and has a lonely life, until he finds the chance to come back for a rematch. However, this time, *Kalunde* did not do so. Instead, he pant-grunted to *Fanana* and strengthened his bond with his bitter enemy. *Fanana* and *Kalunde* apparently became close friends. They often travelled together, pant-hooted together, and did communal 'policing' or 'pacifying intervention' (Boehm 1994), that is, neutral interventions in female quarrels, by just charging past them without supporting either party. They also shared meat so long as they were close to each other. Since *Fanana* as the alpha male was more often the possessor of a carcass, this was *Kalunde*'s gain.

Kalunde, at 37 years old, was lower-ranking than most other adult males in 'basic rank' (Kawai 1958) or 'formal rank' (de Waal 1982) – rank based on a one-on-one match-up. However, to casual observers, he appeared to be the second-ranking male, as long as *Fanana* was not far away. When *Kalunde* was separated from *Fanana*, he screamed incessantly, but without pant-grunting to younger, stronger males when he was challenged by them. At these times, *Fanana* appeared to rescue him and *Kalunde* ceased to scream; moreover, *Kalunde* often counter-charged at the young attacker with the alpha male at his back.

The whole scenario appeared to confer benefits to *Kalunde* rather than to *Fanana*. Why did *Fanana* help *Kalunde*? We may recollect the case of *Kasonta* and *Kamemanfu*. Even if an old male such as *Kamemanfu* is not strong, his existence as a coalition partner was helpful to the maintenance of the alpha position, and the fact that there were only three adult males made the contribution of the old third male even more remarkable. We have witnessed such an arrangement so far only once before *Kasonta*'s case: *Ntologi* and his age-mates, *Bakali* and *Musa*.

Moreover, this was not the first time *Kalunde* assumed this strategy. After *Nsaba* took over alpha status from *Ntologi* in 1995, *Kalunde* became the intimate coalition partner of *Nsaba* and maintained a high-ranking status. *Kalunde* always behaved as if he were a high-ranking male in power politics, both before attaining the alpha position and after losing it. This contrasts remarkably with other alpha males such as *Kasonta*, *Kajugi*, and *Ntologi*, who did not remain in the group once they lost the alpha position. Probably, this was related to *Kalunde*'s old age when he lost status. At his age, a rematch would not have been a feasible strategy, although there was the exceptional case of *Ntologi*.

8.8.3 *Alofu*'s power takeover

Fanana's reign continued until 2003. Unfortunately, no one observed the process of the next power takeover. According to Hitonaru Nishie (2004a, 2004b), *Fanana* suddenly disappeared at the end of November 2003, and was later sometimes seen roaming alone or with a particular adult male, *Masudi* or *Bonobo*, apart from other M-group chimpanzees. Given his absence, the second-ranking male, *Alofu*, automatically ascended to the alpha position. I say 'automatically', but it was more likely that *Alofu* defeated *Fanana* when no observer was present.

In August 2004, Nishie and I were lucky enough to observe an encounter of *Fanana* with *Alofu* and other adult males in the Kasiha valley. To make a long story short, *Fanana* was surrounded by males, escaped to climb up a fig tree, and was penned up in the treetop. *Alofu* was the first to approach *Fanana*, who pant-grunted to *Alofu* with a maximum grin. *Masudi* and a young adult male, *Primus*, quickly climbed the fig tree, rushed at *Fanana*, and also received pant-grunting from him. Only a young adult male, *Darwin*, and an adult female, *Totzy*, approached and pant-grunted to *Fanana*.

Alofu groomed earnestly with *Fanana* in the treetops, while other adult males, *Kalunde*, *Bonobo*, *Pim*, and *Carter*, repeatedly performed charging displays on the ground. Finally, *Alofu* invited *Fanana* to descend the tree together: he repeatedly climbed down a little bit and then glanced at *Fanana*, appearing to wait for *Fanana* to follow him. And every time *Fanana* followed a little bit. *Fanana* nearly reached the ground, shadowing *Alofu*. Then, *Kalunde* and other adult males rushed at *Fanana* and made violent charging displays. *Fanana* immediately fled to climb the tree and returned to his original treetop position.

It appeared that *Alofu* welcomed *Fanana* to return to M-group, but most other adult males, with *Kalunde* as the major dissenter, objected to his return. Perhaps they were afraid of the formation of a strong coalition between the current alpha, *Alofu*, and the ex-alpha, *Fanana*, because this might unfavourably influence their relative statuses. Meanwhile, *Fanana* ran away thorough gaps in the canopy and disappeared.

I was strongly impressed by the sudden change in *Kalunde*'s stance to *Fanana*. When *Fanana* was alpha, he was *Kalunde*'s best protector and coalition partner. Now, he began to select the current alpha, *Alofu*, as his best coalition partner, but he seemed to wish that *Alofu* would not become too strong, so that he would not cease to have any political influence on *Alofu*. This redundancy would imply his political death in his age-class.

It is pertinent to point out here that an overthrown alpha male returns to the group only when he continues to communicate with a part of the 'adult male club' during his days as a fallen power (Uehara *et al.* 1993). This linkage may be a necessary but not always sufficient condition. *Fanana* was sometimes seen to travel with an adult male, either *Masudi* or *Bonobo*, after losing alpha rank, but he never succeeded in returning to his original top position.

8.8.4 *Pim*'s takeover

The latest observation of a power takeover in M-group occurred in 2007, three years before this writing. Nineteen-year-old *Pim* challenged 25-year-old *Alofu*. The takeover seemed to begin in June and finally ended in August, although temporary conflict continued for several months afterward. Luckily, Agumi Inaba and I arrived at Mahale early in August 2007. Although my stay was brief, Agumi stayed long enough to study the process (Inaba 2009), and she will publish a detailed paper on this topic later. Moreover, Miho Nakamura shot a video-tape focusing on the power takeover. Here, I give only a summary.

In June, *Pim*, who usually pant-grunted to *Alofu*, ceased to do so. In August, *Pim* began to threaten or attack *Alofu*, alone. Although *Alofu* occasionally succeeded in enlisting support from other adult males, including *Kalunde*, *Pim* continued to fight on his own. Finally, *Alofu* pant-grunted to *Pim* on 17 August. *Kalunde* initially assumed the policy of 'support stronger', or support the current alpha, *Alofu*, until he later changed his tactics and began to threaten and even attack *Alofu* in the presence of *Pim*. Although even after this incident, *Kalunde* occasionally changed sides, his politicking was eventually inconsequential. Although both *Pim*'s mother, *Fatuma*, and *Alofu*'s mother, *Wakusi*, were still alive, they never intervened in their sons' fighting. Fighting during this takeover basically occurred on a one-on-one basis. Although *Alofu* was not old, he did not leave M-group, but continued to stay there as the second-ranking adult male. Meanwhile, ex-alpha male, *Fanana*, returned to the group, and thus *Pim* currently lives with as many as three ex-alpha males (*Kalunde*, *Fanana*, and *Alofu*).

I would like to note *Kalunde*'s attitude after losing alpha status. He supported the new alpha males in turn, but sometimes ostentatiously showed-off intimate relationships with beta males. He continued such a 'harassing policy' during three power takeover events. Thus, he enjoyed apparently high-ranking status throughout his old age, although he was really in the lowest rank on a one-on-one basis.

8.9 MEAT SHARING AND CHIMPANZEE POLITICS

8.9.1 Sharing with relatives and non-relatives

In the previous section (Section 8.4.2), I briefly explained meat sharing by the alpha male *Ntologi*. Here, I discuss the implications of meat sharing from a broader perspective. Food sharing has been regarded by anthropologists as a uniquely human characteristic for a long time. However, as many animals feed infants or fledglings, human uniqueness in food sharing may lie in sharing food not only with immature relatives, which is likely to be explained by kin selection, but also with unrelated adult members of the group.

It is true that 'courtship feeding', in which a male gives food such as a fish to his female before mating, is seen among many kinds of birds such as gulls, kingfishers, and hornbills. In insects, a female of the Order Mecoptera permits copulation with a male that brings her a large prey (Thornhill 1976). A male mantis allows his consort to devour his body from the head down during copulation. After decapitation, transport of the semen goes more smoothly (Dawkins 1976) – this is the ultimate stage of courtship food sharing!

Thus, in these animals, food sharing between adults is limited to couples or to mates. In general, food sharing in animals is confined to parent–offspring relations and sex partners. Food given by the male to the female partner, as well as foods given by the father to an infant, contribute to the nourishment of the male's offspring most of the time. Therefore, these types of behaviour contribute to the maximisation of one's genes and are thus explained by kin selection.

Consequently, it is important to look for the origins of food sharing among unrelated same-sex adults, in which interactions are not directly explained by kin selection. This is why the sharing of meat among adult male chimpanzees has attracted so much attention.

To repeat, except for humans, food sharing among chimpanzees is different from most other primates in that sharing among unrelated adults is commonly observed, in addition to parent–offspring pairs. Food that is shared between adult chimpanzees is large in size and thus can be divided. Such food includes, for example, sugarcane (Nishida 1970), the meat of vertebrates, large fruits and green leaves on plenty of branches. At Taï and Ngogo, large fruits of *Treculia*,[1] as heavy as 10–30 kg, are shared. At Bossou, papaya fruit is shared (Ohashi 2007).

Under natural circumstances, sharing among adults rarely involves vegetable foods, but is confined mostly to meat sharing. The distributors are usually adult males. This is mainly because males are the only ones who can get their hands on large volumes of meat, but this is not the only reason. Males are far more generous than females.

So, are males generally thrilled to be sharing food? Well, their conduct certainly shows otherwise, most of the time. First, when a male has possession of food, he hides away when his associates come near. When the kill is meagre in size, he holds it in his mouth, making off in any direction. The possessor of the meat usually flies up a tree and attempts to have it all to himself. This is for the basic purpose of cutting out individuals in the 'meat-eating cluster'. The number of seats around the banquet table is scarcer when perched on a limb in

[1] Big fruits of *Treculia* are eaten and shared by chimpanzees at Taï and Ngogo (Nishida, pers. obs.)

Fig. 8.32 *Ntologi* and a meat-eating cluster (courtesy of ANC Productions).

contrast to being seated on the ground (Fig. 8.32). When there is nowhere to run, the carrier of the meat rattles a branch, flails the catch around, barks, and intimidates others. Also, the number one rule of chimpanzees' sharing is 'share the smallest and the most unpalatable piece'. I discovered this rule long ago when K-group chimpanzees shared sugarcane at the feeding ground (Nishida 1973a).

However, when the kill is big, rather than try to monopolise it, it is best for the carrier to ration it out. No single male can wolf down a 10 kg piece of meat in one sitting, and of course, there is no freezer in which to store it. Trying to stop a hungry mob from getting its share would be a waste of time and energy (Wrangham 1975; Gilby 2006). Sharing part of the kill is much easier for the carrier of the meat, rather than monopolising a big slab of flesh that he is not going to eat anyway. Furthermore, it is surely no fun having a faction of the crew breathing down your neck, seeking their share of the flesh.

Does this mean that the carrier of the meat shares only with those who stubbornly pester him over the kill? (Gilby 2006) It may appear to be this way, but as a matter of fact, only very special members of the group pester the meat possessor, as far as I have seen. It appears that the beggar has his or her reasons for so resolutely seeking meat, as other chimpanzees never approach the meat possessor from the outset. So why do such beggars insist so strongly on having a share? The hypothesis that chimpanzees who beg most strongly obtain meat does not explain this (Gilby

2006). If the beggar contributed to the success of hunting, he or she would have reason to require a share. Although this was asserted by Christophe Boesch in his study of Taï chimpanzees (Boesch 1994), it is not the case at Mahale most of the time.

The final possessor of the meat is an individual of great power, particularly the alpha male. During the first ten years that *Ntologi* occupied the alpha rank, we discovered that he had a 'favourites list' of those lucky few who partook in his meat feasts (see Section 8.4.2). Among them were his allies, his mother *Wanaguma*, elderly males and females, oestrous females, and the sex partner he had shown possessive behaviour towards (the most likely candidate to bear his offspring) (Nishida *et al.* 1992).

We have collected data on the sharing behaviour of alpha males in more recent years. Although these data are still being analysed, an interesting trend is emerging. Alpha males succeeding *Ntologi*, such as *Nsaba*, *Kalunde*, *Fanana*, and *Alofu*, have tended to share meat with their most important coalition partners, each of whom might become a supporter of the most formidable rival, according to circumstances. *Nsaba*, *Fanana*, and *Alofu* selected *Kalunde*, the oldest male, as their most important ally. *Kalunde* selected *Dogura*, the third-ranking male. Accordingly, my hypothesis seems to be justified.

The idea that meat sharing shows aspects of reciprocal exchange has been supported by John Mitani and David Watts for the male chimpanzees of Ngogo, Kibale forest (Mitani & Watts 2001).

8.9.2 Highly respected elders

Sharing meat with allies, mothers, or 'wives' is perfectly understandable, but why share with elderly males and females? For elderly males at least, it is because of their political pull.

No matter how feeble an elderly male is in a fight, he is well-respected. They are serene, and because they are rich in experience of all kinds of things, most members gather at their sides. Even when the party is deciding in which direction to roam, they typically first watch the direction the elders take. The high-ranking males even argue over who will groom the elderly. Meat sharing is a way to kiss-up to the elders. Even alpha males such as *Kasonta* and *Sankaku*, as described above, have no hang-ups about parting with a banana or stick of sugar-cane, if it is going to an elderly male.

But I could never put a finger on why alpha males are always so generous in sharing with elderly females. Recently, I found the answer for this when we discovered that old females are fertile, having no menopause (Nishida *et al.* 2003; Emery-Thompson *et al.* 2007). They are given meat, as are other younger mothers, because they too give birth. In any case, it is fair to say that chimpanzees deeply respect the elderly, despite their low-ranking status.

One thing that really piques my interest is that the alpha male never shares meat with the contending rival male (most often, the second-ranking male). *Ntologi* never gave meat to second-ranking *Bakali* in his heyday, but as *Bakali* became older and his status fell, *Ntologi* began to give him meat, as described above. At one point, under no circumstances did the then-alpha male *Kalunde* ever give meat to his rival *Nsaba*. At that time, he often shared meat with *Shike*, who was his most important coalition partner, although he was second-ranking (see Section 8.5).

Most recently, we have been investigating the ways in which the meat recipient 'pays' for his share. Human society is brimming with 'reciprocal altruism' or altruistic behaviours incorporating payback: 'silent trade', pyramid scheme businesses, gifts, trading, taxes, banking, and insurance. It is safe to say that this reciprocity is the greatest distinguishing characteristic of social behaviour in humans. We are exploring the origins of this behaviour through meat sharing among chimpanzees.

8.10 TURNOVER MECHANISM OF POWER

Summing up, for more than 40 years, we have witnessed the succession and resignation of three alpha males in the rather small K-group and eight alpha males in the larger M-group. We can expect male chimpanzees to exploit all that they have, in their social brains in particular, during the very rare opportunities to compete for a power takeover. This is why I am particularly interested in alpha males and the power turnover mechanism.

8.10.1 How do male chimpanzees attain the alpha status?

We have recorded the ages at which males gain alpha status, methods of gaining alpha status, ages at losing alpha status, and tenure or periods of occupying the status. Two adult males attained the status twice in separate periods (Table 8.1). A power takeover event is defined as prolonged fighting between the alpha male and one of the

Table 8.1 *Alpha males of K- and M-groups at Mahale in chronological order*

Name of alpha male	Group	Presence/absence of mother when gaining alpha	Year of birth	Date gaining alpha status	Age (years) on gaining alpha status	Method of gaining alpha status	Date losing alpha status	Age (years) on losing alpha status	Tenure (years)
Kasonta	K	Unknown	*c.*1940	Before May 1966	<26	Unknown	March 1976	36*	≥10
Sobongo	K	Present	*c.*1958	March 1976	18	Alone	1979	21*	3
Kamemanfu	K	Absent	*c.*1940	1979	39	Previous alpha died	1983	43*	4
Kajugi	M	Absent	*c.*1955	1976	21	Unknown	July 1979	24*	3
Ntologi	M	Present	*c.*1955	July 1979	24	Unknown	March 1991	36*	12
Kalunde	M	Absent	*c.*1963	March 1991	28	Coalition	December 1991	29*	0.8
*Ntologi***	M	Present	*c.*1955	February 1992	37	Alone	April 1995	40*	3
Nsaba	M	Absent	1973	April 1995	22	Coalition	January 1997	23	1.8
*Kalunde***	M	Absent	*c.*1963	January 1997	33	Previous alpha died	November 1997	34*	0.9
Fanana	M	Present	*c.*1978	November 1997	19	Alone	November 2003	25*	6
Alofu	M	Present	1982	November 2003	21	Unknown	July 2007	25	3.7
Pim	M	Present	1988	July 2007	19	Alone	Not yet known	Not yet known	Not yet known

Notes

* presumed age

** Two males became alpha twice

high-ranking male contenders, continuing for more than one day. I summarise below how adult males acquire, maintain, lose, and sometimes regain alpha status.

Young adult males may launch a solo challenge for alpha status, and the current alpha may resist such a coup d'etat by mounting a counter-attack, or he may enlist the support of a third party. The current alpha is in a better position than a young challenger to gain supporters, because a third party tends to support the stronger party in what is known as 'winner support'. Furthermore, third parties have a tendency to maintain the status quo, and in any case, the alpha male has cultivated friendships with other males for a longer period than have young challengers.

A young challenging male should not run the risk of severe wounds, or worse, losing his life, because he has not yet produced a lot of offspring. The young adult male avoids risking his higher reproductive value. On the other hand, as the current alpha male has already fathered offspring, he is expected to fight more fiercely. Compared with the young challenger, he has nothing to lose by going all-out. Male–male competition sometimes results in the demise of one party.

8.10.2 Why do challengers win battles?

When an alpha male is challenged for his status by a younger male, the younger male almost always wins the contest, so far as we have observed. Why is this so?

Fighting is likely to be initiated by the non-alpha male. This is probably because the younger one mounts a challenge only when he is confident that he is stronger than the current alpha male. He can select the best opportunity (or age) to make his move, because only he knows when he is in the best condition. The alpha male becomes weaker as he ages, and he cannot pick the opportunity to fight, because he is already the alpha male. Thus, when the challenge occurs, the most favourable time for the alpha male has already passed. Moreover, the alpha does not have a strong reason to pick a fight with a rival because he can already mate with oestrous females as he wishes. As the proverb says, 'The rich do not fight.' The haves should not risk their property by fighting; the have-nots, however, should be expected to fight, because they have nothing at risk if they lose. Thus, the alpha male will not start a battle against the rival.

Of course, he can display against and threaten any rival male anytime he wants. However, as he becomes older, the potential rivals increase in number and displaying becomes more costly and risky. Therefore, a takeover of the alpha status is most likely to be successful. As a matter of fact, takeover trials were observed ten times, and on eight of these occasions takeover was successful.

Methods of takeover depend on the ages of the challengers. Younger males and older ones use very different tactics for takeovers (Table 8.1).

(1) Solo attack is more likely by young males. Those who make solo attacks are the youngest challengers, at around 19 years old.
(2) Attack with a coalition partner is more likely to be used by older males. Those who attack with a coalition partner are a bit older, at around 25 years old.
(3) Automatic promotion is expected more often for older males. From our observations, those who automatically rise in position, after the previous alpha's disappearance, are much older, at around 36.
(4) A solo rematch with the aid of an affinitive network is more likely to be launched by older males. Indeed, those who returned alone but won a rematch by taking advantage of a previous social network are even older, at about 39 years old.

8.10.3 Options open to ex-alpha males after losing alpha status

The first option is living as an outsider, while occasionally picking up an oestrous female. This is the option taken by middle-aged males. Another option is staging a return match and again seizing alpha status. This is the option taken by macho males.

The third and fourth options involve remaining in the group. The third option is to remain in the group and keep a high rank under the protection of the new alpha male. This is an option taken by old political males or relatively young middle-aged males. The fourth option is to remain in the group and be content with subordinate status. This is the option taken by middle-aged or old ex-alpha males.

8.10.4 The basis of alliance formation in adult males

What is the basis for cooperation? Do cooperative partners have special genetic relationships? Considering the alliances between K-group's males, *Kasonta* vs *Kamemanfu* and *Sobongo* vs *Kamemanfu*, genetic reasons are unlikely from the viewpoints of body size and facial characteristics (Nishida 1983a; Nishida & Hiraiwa-Hasegawa 1987).

We have had three cases of maternal brothers who survived after weaning. However, in only one case did both brothers reach maturity. Although they played and associated together often when they were immature, they were not seen to fight cooperatively against a common male rival. We have insufficient DNA data on paternal sibships. We know of only one case of an adult father–adult son relationship from Eiji Inoue's DNA study: *Kalunde* and *Cadmus*. However, we have noticed neither intimate social relationships nor a close association between them. None of my colleagues and assistants had predicted that *Kalunde* would be *Cadmus*' father before DNA technology was introduced.

At Ngogo, Kibale forest, 'paternal brothers do not selectively affiliate and cooperate'. In fact, 'the majority of highly affiliative and cooperative dyads are actually unrelated or distantly related' (Langergraber *et al.* 2007). The situation at Mahale seems to be similar to Ngogo, although more study is needed.

9

Culture

9.1 CHIMPANZEE CULTURE

Human behavioural patterns have some genetic origins, but most, if not all, of them are culturally modified after birth. This also seems to be the case for wild chimpanzees. This finding is one of my most important conclusions after 45 years of research. Humans are not unique in being cultural. Here I sketch out how I reached this conclusion.

I have shown examples of local differences in behavioural patterns in each chapter, ranging across diet, feeding techniques, tools, gestural and vocal communication. Now I give a glimpse of culture in general and then introduce examples of innovation and their spread; that is, fashions and traditions of behavioural patterns and social learning processes, and finally I discuss comparisons with human cultures.

Instinct and learning are two types of behavioural adaptation, although they cannot always be disentangled. In general, the former is employed more by invertebrates and the latter by vertebrates. The former produces quick but rigid behavioural responses while the latter shows slow but flexible responses. Long-lived animals such as primates can afford to have slow responses, given their relatively large body sizes. Larger animals can survive longer without food, so they can tolerate slower responses. Flexible responses are especially appropriate for dealing with complex problems with which animals must cope.

Long-lived animals learn many things in their social and natural environment. Learning may be individual or social. Culture is information obtained by social learning and is shared in common by members of the group, which may be called tradition, if it is passed down from generation to generation. Group-living animals, such as most primates, gain benefits indirectly from culture without risking direct experience (Kummer 1971).

Behavioural patterns to which this definition applies can be found in chimpanzees as well as orangutans, macaques, capuchin monkeys, cetaceans, birds, and many other vertebrates. However, investigations so far have suggested that human and chimpanzee cultures are much more deeply rooted in the species genome. Why do we care about the presence or absence of culture in animals? One reason is that many scholars believe that culture represents the fundamental difference between humans and other animals. This so-called qualitative difference seems to have been used to justify discrimination against non-human creatures. I believe that this anthropocentrism has brought on the global environmental crisis of ecosystem deterioration, including destruction of tropical rain forests and coral reefs and the extinction of many creatures.

9.2 INNOVATIONS THAT SPREAD

In each preceding chapter, I have illustrated some examples of local differences between chimpanzee study sites. Here, I highlight some innovative behavioural patterns that developed among the chimpanzees of M-group and K-group. I operationally define 'innovation' as a behavioural pattern that Mahale researchers never saw before (Nishida *et al.* 2009).The selection of the year 1981 is arbitrary, but 15 years after the beginning of research in 1965 seems long enough for us to notice any common cultural behavioural patterns. Some examples, such as 'habituation to humans', are exceptions of this definition because the transformation of the apes' behaviour was obvious.

9.2.1 Habituation to humans

Mahale chimpanzees' attitudes to humans clearly are learned behavioural patterns, because the apes that fled in the beginning when they first met humans are now completely comfortable with us at close range. We have noticed an intriguing recent tendency: newly immigrated females are habituated to human observers much more quickly than were their predecessors.

It took many years to habituate the chimpanzees to human observers. Some females did not allow humans to approach within 10 m, even after ten years of research. However, young female immigrants recently took only a few weeks to allow observers to approach to within 10 m. In 2002 an adolescent immigrant, *Qanat*, tolerated me to be as close as 5 m within 12 days of her immigration. In 2005 another adolescent immigrant, *Yuri*, tolerated Michio Nakamura to be as close as 3 m on the second day of her immigration (Nakamura & Itoh 2005)!

The apparent acceleration of habituation probably occurred because the newcomers took their cue from the relaxed attitudes to human observers of the already-habituated chimpanzees (Nishida 1987).

9.2.2 Eating cultivated items

Before I began my research in 1965, the chimpanzees of Kasoje ate only the piths (but not the fruit) of banana and piths of maize, among the many cultivated plants available. However, a few chimpanzees began to eat the fruits of guava and mango trees in 1981, and lemons in 1982, and these food habits spread quickly to other members of M-group (Takahata *et al.* 1986). The fruits were not provided at the feeding place, but the chimpanzees voluntarily began to eat from the standing trees near the old village sites.

Why did they not begin to eat these fruits much earlier? This may be because these fruit trees were planted so close to the villages that the chimpanzees were afraid of the villagers. The human residents were moved outside of the proposed national park area by 1975, and as a result, the apes were able to freely visit abandoned village sites and so had opportunities to taste the fruits. However, this does not explain why they did not begin to eat the fruits of oil palms, which are eaten by yellow baboons, vervet monkeys, and red-tailed monkeys living within M-group's range, and which are the most important food of Gombe chimpanzees.

9.2.3 Wipe and rub

Chimpanzees at Gombe use 'leaf napkins' to wipe sticky substances, such as faeces, blood, seminal fluid, and fruit juice from their bodies (Goodall 1986). So, their wiping behaviour is customary.

However, I have so rarely seen such wiping behaviour that I can remember almost every case. For example, an adult male named *Nsaba* once used a leaf to wipe his penis clean after copulating.

Another adult male, *Kalunde*, wiped away faeces deposited by another chimpanzee on his back. An adult female, *Wakusi*, wiped from her back urine dribbled on her by her infant daughter. An adult male, *Fanana*, several times wiped lemon juice from his mouth with a large leaf (Nishida *et al.* 2009).

When the perimeter of a chimpanzee's mouth is sticky with sap from eating *Parkia* or *Saba* fruit, she or he cleans the face or muzzle by rubbing it against tree branches, some shrubs, or stems of grasses.

When an ape has trod on some faeces, he or she repeatedly rubs the hands and feet on dry leaves or the ground. Such muzzle or hand rubbing was rare at Mahale, but since 1998, such hygienic rubbing has increased when eating lemon fruits, and by 2003, 29 chimpanzees showed this pattern (Corp *et al.* 2009). (This common behavioural pattern is not classed as tool use, because the object used has not been detached from the environment.[1])

9.2.4 Belly slap

In 1999 an eight-year-old male, *Cadmus*, was first seen to slap his abdomen with his palm while hanging from a tree, supported by the other hand. In October 2000 a five-year-old male, *Xmas*, began to do the same thing, which became his favourite activity (Fig. 9.1). In the same month, a two-year-old male, *Caesar*, also joined in. First performed as solo play, by 2003 the pattern was used often in play solicitation to another youngster, and in 2004 it began to be used in intimidation displays to adolescent females and juvenile males. In 2003 three-year-old *Xantippe* slapped her belly while hanging from a tree; she probably learned this behaviour from *Xmas*, her brother. In 2004 an eight-year-old male, *Michio*, began to slap his belly, which he probably learned from his

Fig. 9.1 A juvenile male, *Xmas*, slaps his belly in a tree.

[1] See Beck (1980) for definition of a tool.

close playmate, *Xmas*. In 2005 *Cadmus*, then 14 years old, still slapped his belly in play solicitation or intimidation displays. In August 2007, both *Cadmus* and *Xmas* still showed belly slapping. As all were playmates, *Xmas* likely learned the pattern from *Cadmus* (Nishida *et al.* 2009).

9.2.5 Water play and sponge use

As mentioned briefly in Chapter 4, leaf sponging was a tradition discovered long ago among the chimpanzees of Gombe but was not seen in Mahale's chimpanzees. In 1985, two adult female immigrants from K-group engaged in leaf sponging a few times, but no such observations were made again until 2000, when two juvenile females began to use leaf sponges, and by 2005, 11 youngsters did so. These youngsters had no chance to see the leaf sponging of 1985, so their actions were innovations, probably spread by observational learning (Matsusaka *et al.* 2006). 'Leaf spoon' was seen in a juvenile male for the first time in 2000, a juvenile female in 2001, and another juvenile male in 2002, suggesting another case of transmission among youngsters by observational learning (Nishida *et al.* 2009).

9.2.6 Innovations forgotten and rediscovered

It is unlikely that only during our research were such new discoveries made by chimpanzees for the first time in the history of M-group. Why did we not see such water play for 15 years, but then suddenly begin to see as many as 11 youngsters engage in it?

I have tried to draw a scenario for such incidents. Probably, youngsters occasionally 'invent' play such as leaf sponging without being observed by researchers. At some point, a behavioural pattern spreads to others by observational learning and emerges as a fashion or temporary custom. As a result, researchers notice the innovation. The invented pattern continues for a decade or so, but then disappears, perhaps because of the lack of an appropriate cohort of play-mates. For example, an epidemic might kill most of the young infants who would otherwise inherit the fashion from the older generation, or kill most of the older infants or juveniles who have maintained the fashion. Later, the habit might reappear as an innovative pattern turned fashion, after decades. I have come to think of the process like this, because behavioural patterns such as leaf sponge and leaf spoon are simple tool-use in play, and youngsters could hit upon this idea at any time, as with 'potato washing' in macaques. In chimpanzees, the reason for such a fashion failing to become established as a long-standing tradition may be the lack of a large cohort of youngsters. As unit-groups usually number 40–60 members, the occasional absence of cohorts of youngsters for either transmitting or inheriting cultures may be typical in nature (Nishida *et al.* 2009). Since water play and belly slap are juvenile culture, they appear as fashion rather than tradition.

In Chapter 4 I introduced 'leaf-pile pulling' as an example of innovative play. This pattern was first seen in 1989, but was not seen again until 1999. In 1999, within a year, 15 immature chimpanzees engaged in such play. It seems likely that invention of the play pattern occurred twice, around 1989 and 1999, but that the 'first' invention was forgotten, while the second caught on to become a fashion in the group.

9.3 INNOVATIONS THAT FAILED TO SPREAD

Unlike cohorts of youngsters, cohorts of adults cannot be missing; if they disappeared, the group would go extinct! So, the 'loss of cohort' hypothesis may not apply to adult innovations so long as the unit-group continues.

9.3.1 Probe to clear blocked nasal passage

In 1992, from August to October, a flu-like epidemic was going around M-group. About 70 per cent of the group members got sick and among them was an adult male, *Kalunde*, who had a bad case of stuffed-up nose. At that time, he took a thin probe and began to stick it slowly up his nostril. '*Aatchoo!*' Out came a great sneeze and the snot flew! He then repeated this same action with his finger, sopped up the snot, put it in his mouth, and ate it. There is no doubt that *Kalunde* intentionally induced a sneeze by using a tool, because he did the same thing four times, and when a sneeze would not come out of one nostril, he tried probing the other (Fig. 9.2a, b).

How did *Kalunde* arrive at this kind of insight to implement such tool-using behaviour? Perhaps when he was walking around the forest, a branch accidentally poked his nostril and he sneezed. Or, when he was gathering ants, his nose felt itchy and without thinking, he used the probe already in his hand to scratch that itch. By doing so, he induced a sneeze and as his nose then felt relieved, perhaps he learned that when your nose is itchy, inserting a probe into the nostril provides relief (Nishida & Nakamura 1993).

(a)

(b)

Fig. 9.2a, b *Kalunde* pushes a probe into his nostril and sneezes (courtesy of Anica Productions).

Why did this pattern fail to spread? Perhaps too few chimpanzees saw *Kalunde*'s behaviour when they were catching colds. Or, perhaps, many chimpanzees resist pushing a straight probe into a sensitive part of the body as they consider it harmful.

9.3.2 Algae eating

A kind of freshwater algae grows in the larger streams of Mahale, but until 1997 chimpanzees were not seen to enter any stream to eat it.

Then, in September and October, Tetsuya Sakamaki saw an adolescent female, *Sally*, who had immigrated into M-group in January, enter a stream and eat algae. Some M-group chimpanzees intently watched her do so (Sakamaki 1998). However, although more than ten years have passed, no other chimpanzee has been seen to eat algae, and *Sally* has not been seen to do it again since 1998. It is as if *Sally* gave up continuing a custom of her natal group when no other chimpanzee in her adopted group followed suit. That leaves just one idiosyncratic case of algae eating over 40 years of research (Nishida *et al.* 2009). Although plenty of edible vegetation occurs in a swampy environment, M-group chimpanzees gather it above the water's surface, but typically not below it. This may explain why algae eating did not spread to other members.

9.4 WHAT SPREAD AND DID NOT SPREAD

As described above, there are cases when an innovative behavioural pattern introduced to M-group did not propagate to other individuals. Why do some patterns spread and others not?

9.4.1 Enculturation versus propagation

Information flow from many individuals to a single individual ('many-to-one') is easy, but flow from a single individual to many individuals ('one-to-many') may be much more difficult. The former process is called 'enculturation' or socialisation, and the latter 'propagation' (Nishida 1987). Thus, newly immigrated females habituate quickly to human observers as enculturation. Neither nasal probe nor algae eating spread, primarily because it was one-to-many. At Gombe, however, a female chimpanzee with a technique for carpenter ant fishing emigrated from the Mitumba to the Kasakela group. Later, some chimpanzees of the Kasakela group began to fish for carpenter ants (Goodall, pers. comm. 1 November 2008). This may be the first confirmed evidence that a new pattern imported by an immigrant female propagated to the new group, albeit at a slow tempo.

9.4.2 Imitation

Although it is obvious that wild chimpanzees obtain some information from other members of the society, it is hard to elucidate the mechanism by which the information is transmitted socially.

Infant chimpanzees seem to begin ant fishing by watching their mother's actions and manipulating her probes. Only after lots of trial and error do they begin to fish for ants, at about the age of three years. This process could be called 'local enhancement' (Nishida 1987) (*sensu* Thorpe 1963).

'Imitation' is operationally defined here as an individual performing an act within a minute after watching another individual doing the same act. During the study period of 1999–2004, I recorded 44 examples of 'imitation' for youngsters (1–5 years old). Youngsters were influenced by the behaviour of other unrelated immature individuals on 17 occasions (38.6 per cent), by their mothers on 14 occasions (31.8 per cent), by unrelated adults on 7 occasions (15.9 per cent), and by older siblings on 6 occasions (13.6 per cent). Most (36) of the 44 examples were due to 'local enhancement'; or, this possibility at least could not be excluded. These cases cannot be called 'true imitation' (*sensu* Thorpe 1963). For example, a two year-old female, *Xantippe*, showed leaf-clipping eight times in 42 seconds after her seven-year-old brother showed the display. However, she had already showed this pattern when she was only one year old. Thus, her behaviour would be categorised as a result of 'social facilitation' (*sensu* Thorpe 1963). In only eight cases did a youngster's first-observed performance accord with his or her imitative behaviour. For example, a one-year-old male, *Ichiro*, bent a shrub immediately after a three-year-old male's shrub-bend performance.

In nature, it is impossible to determine whether or not a behavioural pattern is 'true imitation', because we cannot monitor all of the behavioural output of an individual. In spite of this, I argue that 'imitation' is an important constituent of chimpanzee observational learning. Unless this is presumed, it is impossible to explain why young chimpanzees do an act immediately after an elder one did the same. Young chimpanzees are said to show some imitative facial expressions in captivity. There seem to be no reports of monkeys such as macaques showing 'imitative patterns', and their lack of ability to imitate in captivity is a notable contrast to chimpanzees.

9.5 HUMAN AND CHIMPANZEE CULTURES COMPARED

With regard to culture, a huge discrepancy exists between humans and chimpanzees. It is a question of education. Chimpanzees hardly ever teach in a proactive manner.

Teaching can be divided into 'encouragement' and 'discouragement' (Nishida 1987). I have never seen chimpanzees apply encouragement in their teaching. Christophe Boesch once wrote about a mother chimpanzee teaching her child to crack nuts with a stone. This is the only anecdotal example of encouragement reported in the education of a wild chimpanzee (Boesch 1991b). Even in captivity, little evidence of teaching has emerged. At the Yerkes Primate Research Center, there was a report of a mother encouraging her baby to walk, using body movement and vocalisations (Yerkes 1943). Felid mothers such as cheetahs bring a newly caught prey that is half-dead to their young (Caro & Hauser 1992) as this allows them to start acquiring the art of hunting, this is teaching by encouragement. Thus, humans and animals cannot be distinguished by teaching by encouragement or not. However, felid teaching is limited to hunting and never anything other than this. Human teaching is varied and spans all behavioural domains.

Teaching by discouragement occasionally occurs in chimpanzees. When a baby chimpanzee, brimming with curiosity on its first encounter with an unfamiliar object, edges in closer to it, the mother generally steps in the way. If a baby chimpanzee starts to put into its mouth a leaf that is not part of the group's dietary repertoire, then its mother snatches the leaf from its mouth and tosses it away. Even these cases are extremely rare (Nishida 1983b; and later observations); I have seen it only a few times over 40 years.

9.5.1 Human uniqueness: imitation and the ratchet effect

Recall that *Kalunde* used a probe to stimulate a blocked nasal passage. What intrigues me is that the only chimpanzee in M-group who uses tools in this way is *Kalunde*. The other chimpanzees have seen him do this time and again, yet none of them imitates him. *Nsaba* and others have peeped at *Kalunde*'s behavioural patterns before, and according to Kenji Kawanaka, in the following year about the same time, the flu was going around again, but the only member of M-group making use of a probe to induce sneezing was *Kalunde*. Koichiro Zamma and I also watched him doing the same thing in the early 2000s (Nishida *et al.* 2009).

Comparing chimpanzees to humans, it is apparent that chimpanzees do only the least amount of behavioural imitation. We often hear 'monkey see, monkey do', which is an exaggeration – it should be 'human see, human do'.

Another difference between human and non-human cultures that has often been pointed out is the accumulation of information or what has been recently dubbed the 'ratchet effect' (Tomasello *et al.* 1993). Information in human culture accumulates year by year and generation to generation. Thus, each new genius begins her or his inventiveness from a high starting point. This is unlike a chimpanzee genius who must begin his or her invention anew. Innovation in a chimpanzee group is not rare, but many innovations are easily forgotten, and the same innovation appears repeatedly after several decades. A new behavioural pattern lasts as a fashion for several years, but then disappears. This is because the size of a chimpanzee group is small, and if members of a particular age or sex class are few or absent altogether, then the innovation will not be transmitted to another generation (Nishida *et al.* 2009).

Thus, we can see much progress in human cultural history. Although this appears to be a great difference between ape and human, the ratchet effect seems to be relatively new. Oldowan and Acheullian 'stone age' cultures lasted for more than one million years, each without substantial changes (Schick & Toth 1993). The remarkable ratchet effect is a new phenomenon introduced only after *Homo sapiens* emerged, and, in particular, after the exploitation of fossil energy.

10

Conservation and the future

My research on the wild chimpanzees of Mahale started 45 years ago. I have spent only 12 years in Africa, but almost always I have been preoccupied with the chimpanzees of Mahale, even when in Japan. I have absolutely no regrets about dedicating my whole professional life to research on chimpanzees. The chimpanzees are really worth the time we have spent together, and they are new to me each time I visit Mahale.

In pursuing my work, I have lived in one of the world's most remote places, where people continue to lead a traditional village life. My cottage was more humble than most households, without any of civilisation's amenities, such as electricity, plumbing, gas, newspaper, or telephone. I enjoyed the company of cheerful people, a green forest, wild animals, fresh air, clean natural water, and additive-free food, including fresh lake fish, delicious chicken, and vegetables.

In Japan, I spent 20 years in Tokyo and 25 years in Kyoto. When I began my study, Japan had just started its high economic growth, and now it is one of the most developed countries. I have experienced life at two extremes: a traditional rural society and a highly industrial society. Accordingly, I have seen and lived the contrast between these extremes, an opportunity rarely experienced. Since the traditional society is vanishing quickly, I feel I must leave something to future generations.

10.1 THE SIGNIFICANCE OF CHIMPANZEE RESEARCH

Chimpanzee research is a worthwhile pursuit in its own right. It has value for science, entertainment, conservation, and local economies.

First, I want to reconstruct the society of our last common ancestor. The societies ancestral to humans, chimpanzees, and bonobos shared some common features. The first such feature was a patrilocal system, in which females mated outside of the natal group while males reproduced within it. The second feature, unlike gorillas, was compromise among adult males in considering mutual costs and benefits and cooperation with one another when necessary. This was a multi-male–multi-female society, lacking pair-bonds or strong male–female bonds. Each patrilocal group (social unit) was antagonistic to other groups. Mothers reared children, but they also opportunistically took advantage of the child-rearing assistance of adolescent females, sterile females, and, less often, adolescent males of the same group.

Sometime after the separation from the *Pan* lineage, male–female bonds and human families crystallised within the original multi-male–multi-female society. Some families were monogamous and some polygynous, depending on the strength of the male. Owing to the attachment of females to each male, the hominin ancestor continued the unit-group system as a community based on male cooperation, maintaining the patrilocal system from the last common ancestor (LCA). Late adolescent females left the natal group, immigrated into a neighbouring group, and gave birth to children. Most importantly, however, fathers and mothers retained some connection with daughters who married outside the group. Parents may have visited their daughters, and daughters may have revisited their natal communities. This continuing connection may have been the most important feature of hominin society, as it enabled communities to make alliances with each other through the 'exchange' of females.

The second most important feature of the society of our hominine ancestor was the nearly obligatory mutual assistance in rearing children within the community. Not only immediate relatives but also distant cousins helped to rear one another's children, including even the adoption of distant relatives. Most importantly, this was maintained by tradition underpinned by genetic propensity. I do not extend my speculation on the society of our hominin ancestors beyond this point, because it is sufficient for me to discuss my ideas on the deterioration of modern society.

The child-rearing tradition of mutual assistance, which has continued for one or two million years, recently has collapsed in modern urban society. Although the extent of deterioration varies from culture to culture and from nation to nation, suicide, homicide, abuse of others or oneself, bullying, and psychological disorders have increased among people living in the cities of developed countries. I believe the collapse of child-rearing traditions is the main reason why such tragic phenomena have emerged. People have no relatives nearby, and wives and husbands have no reliable relations except with each other. Without assistance close at hand, young mothers, in particular, are at a loss when difficulties occur in child-rearing. On such occasions, people previously had recourse to aid from relatives in the same community.

The collapse of such helping systems was very recent. Let me give an example from my own lifetime. Fifty years ago, all of my paternal uncles and aunts were still living nearby in the suburbs of Kyoto, and these relatives often met. Every year, at least in the summer and winter holidays, my brothers and I visited my father's birthplace, where my grandfather, my father's eldest brother, and their families were living, and spent many days playing with our cousins. As a further example, in those days, a pregnant Japanese city-dweller typically returned to her home district to obtain help from her own mother and relatives, staying there for more than a few months. Therefore, the community-like system was maintained to some extent, even if relatives were not living in the same community.

I would like to propose the above hypothesis in searching for a solution to the pathologies of modern city-dwelling from the perspective of primatology.

10.2 ECOTOURISM

10.2.1 Disease transmission

Since Goodall first suspected that the chimpanzees of Gombe may have suffered from polio transmitted by people from a nearby town (Goodall 1968), attention has been paid to the possibility of chimpanzees suffering from human-transmitted disease or, more generally, common diseases between humans and other great apes.

At Mahale, we considered this in 1977, when some chimpanzees of K-group showed symptoms of colds at the same time that village people also caught colds (Uehara & Nyundo 1983). In 1993, when nine chimpanzees died of a flu-like disease (Hosaka 1995a), some village people also suffered from flu-like symptoms. At the time, I regretted that we had not taken precautions to avoid transmitting diseases. However, bad events are easily forgotten, especially considering that respiratory diseases usually do not kill chimpanzees.

However, in 2006 another outbreak struck M-group chimpanzees. Five individuals showed the symptoms of a cold and subsequently disappeared. Three other chimpanzees were not seen to suffer from the illness, but as they disappeared during the outbreak, it was suspected that they too were victims. Two adult and two infant bodies or skeletons were recovered. Postmortem examination by a veterinarian showed that one of the infants suffered from pneumonia. This time, we immediately changed our observation methods. We made it a rule to wear face-masks whenever we were watching chimpanzees (Hanamura *et al.* 2006, 2008).

Later, a variant of *Parapneumovirus* that before had been detected only in humans was found in the faecal sample of a sick adult female, *Wakusi*, providing decisive evidence that the flu-like disease came from humans (Kaur *et al.* 2008). Around the same time, in Taï Forest, Ivory Coast, human pathogens were found in the faecal samples of that site's chimpanzees. There, as no ecotourism occurred, pathogens must have originated from researchers (Köndgen *et al.* 2008). So, this possibility must also be considered at Mahale. The Mahale Wildlife Conservation Society (MWCS) decided to provide face-masks to each person seeking to approach wild chimpanzees. Fortunately, since September 2006, when the park management of Mahale National Park and tourism companies consented to require all staff members and tourists to wear face-masks, no chimpanzee has shown serious symptoms of respiratory disease (as of December 2010). Simply wearing a face-mask is not a sufficient method of preventing disease transmission, so we are trying other methods, such as a quarantine period after a new person arrives, in addition to strengthening the current 'chimpanzee viewing rules' that we established 17 years ago after the 1993 epidemic (Table 10.1).

Table 10.1 *Chimpanzee viewing rules for tourists at Mahale*

Proposed by me	Currently practised
1 No sick persons should observe the apes	Coughing heard rarely
2 All persons should wear face-masks	Strictly observed
3 No toilet in the bush	Usually observed
4 No smoking and no perfume in the bush	No smoking, but many female tourists wear perfume
5 Keep 10 m from chimpanzees	Sometimes <5 m
6 Observe chimpanzees while seated	Often standing
7 Two tourist groups per day	Often more than three groups
8 One tourist group consists of six tourists plus one guide	Usually more than two guides
9 One tourist group has a maximum of one hour of observation	Usually observed
10 No tourist should enter the bush	Rarely observed

10.2.2 Ecotourism and stress on chimpanzees

One afternoon in the early 2000s, I was quietly watching several chimpanzees grooming. Some were sitting and others lying leisurely on the ground. Suddenly, all of them sprang up and fled into the trees so quickly that I was taken by surprise, having no idea what was happening. Then four men appeared from behind me, carrying an old woman on a stretcher. The chimpanzees were surprised at the strange sight of a supine tourist carried by bearers.

At other times, I was impressed to see elderly women who could walk only with a cane, even over flat terrain, make every effort to climb a steep slope to catch up with chimpanzees. I appreciate that watching wild chimpanzees is highly valued by Western people, even when such an earnest effort is required. Rich Japanese people should follow this example. However, we should not surprise chimpanzees by introducing such a human custom as bringing people on a stretcher, which is totally alien to wild apes. Stress significantly disturbs the immune system of chimpanzees, so we should be careful not to introduce unpleasant stimuli to their environment. Many tourists go into the bush wearing colourful clothes or smelling of perfume; a female tourist once surprised me when she partly undressed and followed chimpanzees into the bush wearing a bikini-style swimming top!

10.2.3 Ecotourism and community conservation

Ecotourism centred on chimpanzees has pros and cons. As mentioned above, ecotourism has some costs for wild chimpanzees, including stress and disease transmission. Efforts to minimise the risks should be made by all means available. However, the benefits of ecotourism are great. Ecotourism not only provides precious foreign currency to the government, but also stimulates local employment and brings business to local markets. However, local residents who surrender their traditional land to public use should be properly compensated and rewarded. Ecotourism will encourage public awareness of the importance of natural conservation. Even among the village people of Kasoje, only a few had seen chimpanzees close up, and so they did not know how much wild chimpanzees act like humans.

In order to avoid risks, only one tourist company should be authorised to run the tourist business under the supervision of an official agency. Competition among tourist companies brings disastrous effects by encouraging them to offer tourists services such as observations of longer duration and closer proximity at the same price. At Mahale, three tourist companies plus Tanzania National Parks (TANAPA) currently are in the business, and more than 2000 tourists visit the park annually. Consequently, ecotourism nowadays has become nearly ordinary commercial tourism (Nishida & Nakamura 2008; Nakamura & Nishida 2009). There is no booking system supervised by the park management, and we are greatly afraid of what will come to pass if the number of tourists increases.

10.3 KEEPING A PRISTINE ENVIRONMENT

In October 2007, all of the guava trees (*Psidium guajava*) in Mahale Park, from Kasiha to the Kansyana research camp, were cut down by workers of the Frankfurt Zoological Society in collaboration with TANAPA. In response to protests by Michio Nakamura, who was present at the time, they stopped cutting the remaining trees (Nishida 2008a). The reason given for felling the trees was adherence to the policy of exterminating all alien trees planted by humans, which according to TANAPA employees, is stipulated in the draft Mahale Park General Management Plan (Tanzania National Parks 2006).

I have reservations about trying to exterminate all introduced plants because some of them are part of the cultural heritage of the human residents and materials for conservation education, as well as being important food sources for chimpanzees.

10.3.1 Guava trees thrive in the park

Until the late 1970s, there were seven hamlets at Kasoje along Lake Tanganyika in the current national park area, from the south of Lubulungu to the north of Kasiha. When I arrived at Kasoje in 1965, I saw oil palms, hedgerows such as *Jatropha curcas*, and fruit trees such as mango, lemon, orange, papaya, coffee, and banana. On the other hand, guava trees and *Senna* (ornamental shade trees) were introduced to Kasoje from a nearby village in the late 1960s, after I arrived. At Kansyana, lemon trees thrived and made an impressive grove after I threw away food garbage around the camp in the 1960s, without any intention of creating a lemon plantation. After the establishment of the national park in 1985, trees such as banana, coffee, and papaya disappeared rather quickly. Other trees such as mango and oil palm thrived but never extended their distribution beyond the old hamlet sites. Only *Senna* reproduced robustly at the cost of indigenous vegetation, thus becoming a notorious alien, invasive plant (Turner 1996; Nishida 1996b; Lukosi 1997; Wakibara 1998).

What about the guava and lemon trees? Casual observers seemed to think that these trees were also invasive species, but this was never the case. Let me recount an interesting story. Perhaps more than ten years after the guava and lemon trees around Kansyana grew large and began to bear fruit, some chimpanzees began to eat these alien fruits, and this food habit spread rather quickly (Takahata *et al.* 1986). When M-group chimpanzees travel, they usually use vegetation-free paths, unless they need to enter the bush to eat, take a rest, or avoid tourists. So, there are plenty of opportunities for them to disperse the seeds of cultigens along the paths. When it begins to rain in October, it is nearly the season when field assistants start to cut down shrubs, herbs, and grasses growing on the paths. The seeds of cultigens also germinate then, and thus their saplings might also be destined for removal. However, the keen eyes of the Tongwe assistants easily discriminated the guava and lemon seedlings from those of other plants. They almost reflexively avoided removing the seedlings of cultigens, because they were, after all, born farmers, although we never told them to rescue cultigens. Their normal work had been to remove weeds and to nurture cultigens, and they also were trained to avoid cutting the food plants of chimpanzees.

After several years, I noticed a row of guava trees along the J-Road and Route 1, from Kasiha to the Kasiha River, through the Kansyana research camp. I realised that this was collaborative work by chimpanzees and humans! I left this situation as it was and did not tell the field assistants to cut down the guava trees. I made this decision because guava fruit seemed to have become a substantial component of the diet supporting chimpanzee life at that time. June is a lean season of fruits in many years, and guavas were thus a lucky gift to the chimpanzees. Guavas seemed to compensate for the loss of food supply resulting from the invasion of *Senna spectabilis*. Lemon trees were apparently less strong than guava trees in competition with natural vegetation, but they also thrived along the observation path, thanks to the unintentional human intervention. Lemon trees in the research camp were visited by chimpanzees many times, particularly in September 1999, and they appeared to provide one of the most important foods in that season for M-group chimpanzees.

10.3.2 Illusion of a pristine environment

We should have no illusion that 'natural land' exists without the presence of humans. On the contrary, human beings and their ancestors have lived with other creatures since the emergence of such life forms on Earth. Even at Mahale, if you climb Mt Nkungwe you can enjoy a beautiful landscape of woody fern (*Cyathea* sp.) and giant trees such as *Parinari*, *Anthonotha*, and *Croton*, which appear to be ancient and pristine. However, if you look carefully, you will find pieces of clay pots buried near the highest end of many steep valleys. Tongwe people have lived there for at least 130 years (Stanley 1878; Nishida 1990b). Residents evacuated their traditional land, which they had inherited from their ancestors as their most important treasure, and ceded it to the government. Mango trees and oil palms are evidence of their historical presence, together with natural monuments such as huge rocks that they believed harboured the guardian spirits. These are part of the cultural heritage of the Tongwe people.

The Mahale National Park should use this cultural heritage as one of the available teaching materials and attractions for tourists, as well as an expression of respect and gratitude to the original residents. Tourists should be thankful to those who surrendered their land as a national treasure to many other people at the cost of their village, subsistence base, and memorials to ancestral spirits. Moreover, every human with common sense knows that mango trees and oil palms provide good shade, which tourists need in the sunny dry season. The row of guava trees alongside Route 1 would be an interesting resource for teaching about seed-dispersing activities based on both chimpanzee and human intervention. These trees are alien (although the oil palm is of ancient West African origin) but never invasive. Some have suggested that oil palms attracted yellow baboons to inland park areas and thus robbed chimpanzees of some inland food patches (Nishida 1997b). If so, then some of the oil palms from the Kasiha workers' camp to the west of Kansyana could be cut down, but the other oil palm groves should be left intact. I would also like to emphasise that the observation paths should be no more than 1 m wide. I was surprised to see that TANAPA temporary workers sometimes widened paths up to 3 m, cutting down shrubs and woody vines constituting important dietary components, such as *Psychotria peduncularis* and *Ficus urceolaris*, because they had no training about vegetation. Tourist companies welcome wide paths for their convenience, but this comes at the cost of the chimpanzees' subsistence.

10.4 PRESERVATION OF THE CHIMPANZEE

10.4.1 Devastating impact of deforestation

The chimpanzee virtually follows the course of the equator, ranging from Senegal, West Africa, through Central Africa, and all the way to Tanzania, East Africa. A great river, Ugalla River, runs 150 km east of Mahale, marking the eastern boundary of chimpanzee distribution. For living space, the chimpanzee can utilise even dry regions like the savanna, where one may find baobab trees growing (McGrew *et al.* 1981; Teleki 1989), but if the riverside forest does not develop, then the apes will not survive (Kano 1972).

In most African countries, especially in West Africa, the chimpanzee is extinct or limited to a few select regions, and only a fraction of previous populations hang on, and those barely (Kormos *et al.* 2003; Caldecott & Miles 2005). About 200 000 chimpanzees are estimated to live in Africa, although no one actually knows for sure (Butynski 2001). National parks and other agencies, which are better informed of the actual situation, claim that the total chimpanzee population is a mere 15 000 (Wrangham *et al.* 2007)!

The chimpanzee's habitat, which is African tropical forest, has been turned into farmland, deforested for lumber and mining production, and chopped down for firewood, and over the past 50 years, more than half of this forest has been wiped out.

Even today deforestation progresses, with 13 000 km^2 of tropical rain forest disappearing annually. Added to this, chimpanzee mothers are killed and the infants are dragged off to fill the demands of zoos for their exhibits, to become people's pets, and to be used in medical experiments. In the final analysis, since chimpanzees and other living creatures have fallen prey to humans, as long as our species' population continues to swell and we continue to raise our standard of living, their chances of survival are next to nil.

10.4.2 The chimpanzee next door

Economic development and advanced technology will never be the solution to the predicament of chimpanzees. Consider agricultural development, for instance: to alleviate hunger, we clear-cut forests, create farmland, and try to plant high-yield crops. Without doubt, food supply increases temporarily, but with increases in food supply comes an increase in human population, taking us back to square one.

Is our prime purpose of living to feed as much of the human population as possible at the expense of other living creatures? That seems to be the direction modern politics is going. Since natural resources are limited, the economy cannot thrive forever, nor can we sustain an ever-larger population. And if 'leading a healthy and cultural life' is the objective, sooner or later 'humanity's purpose' and 'personal purpose' will lead to a paradox. We are already knee-deep in paradox. In Asia and Africa today, the fact of civilised life is that hundreds of millions of people are suffering from starvation. A healthy and cultural lifestyle in the northern hemisphere is founded on starvation and destruction of nature in the southern hemisphere.

Countries in the northern hemisphere are also destroying their own natural environments; look at the current state of Japan. Although it is totally unnecessary, roads are constantly being built, rivers are being dammed, seashores are being cemented, and land is being reclaimed. The biggest victims are the living beings that have been killed and whose habitats are devastated. Because Japan's economy has thrived mainly through destruction in the guise of construction, as shown by the corruption between general contractors and politicians, the situation is serious.

On another issue, the ethical question of organ transplants from brain-dead donors has sent the media into a frenzy, but in Japan the use of animals for medical research, pharmaceutical production and testing, and various experiments has not raised the least controversy. However, making too sharp a distinction between humans and all other living creatures is clearly and ethically a far more serious issue than a discussion of brain-death.

Humans once discriminated against groups other than their own, and would call only the members of their own tribe 'human'. They would either kill members of another group or use them as if they were livestock, and this was considered a reasonable practice. Slavery ended, apartheid was abolished, and all races finally were recognised as human. Next, in a logical progression, the chimpanzee should be recognised as a 'friend of man'. Though considered separate species, chimpanzees and humans share 99 per cent of their DNA. The time to end the persecution of chimpanzees is long past. Furthermore, a heightened consciousness of the need to prevent the extinction of other living things, at the very least, is absolutely essential.

Humans are a species of animal and we could not survive without exploiting other animals and plant life to some extent. However, exploitation should be moderate, and the simple rationale of 'it is for the good of humankind' does not make the excessive exploitation of animals a pardonable act. Why should some six billion human beings get away with the exploitation of the total world chimpanzee population (a measly 200 000 at the most) for the sake of medical experiments? Even today, some medical professionals exploit animals for what they call a 'noble cause'. Why is only human life considered so precious?

Jane Goodall says that chimpanzees are the bridge linking humans and animals. When we consider the current dire state of the chimpanzee's prospects, the notion that other living creatures have a right to life as well will inevitably sink in. Until we become satisfied with simpler lives and discard our anthropocentrism, revealed by such detestable bromides as 'human life is more precious than the Earth', the only alternative will be the eventual extinction of humankind and all other living organisms.

The days of expansion and development are over. How long are we going to keep talking such nonsense? Why is progress such an imperative? What real progress came with civilisation? The words 'economic growth rate' dance across the page of every daily newspaper, and we live and die by these words. Nobody ever asks, 'Why must we grow and develop?' We are obsessed with the pointless debate over development, without asking the most basic question: what is the meaning of life?

Petroleum and other resources are non-renewable – finite – and renewable resources such as agricultural produce and so on have a maximum yield that is fixed. Therefore, economic growth and a rising population cannot last forever. This is due to the simple fact that the destruction of nature is proportionate to human population size multiplied by chosen standards of living.

In the final analysis, we humans are responsible for coming up with an optimum population and a sustainable 'mode of living'. The latter is a difficult concept to define, but something like 'seeking an end to the destruction of nature before damage becomes irreparable' is appropriate. As a case in point, if Japan alone maintains such a sustainable mode of living, its population (currently 120 million) will not be much higher than 30 million, which is approximately the country's population during the Edo period (1603–1868). Accordingly, the recent trend towards population decline in Japan should be warmly welcomed.

Obviously, this problem cannot be resolved by Japan alone. Globalism makes it hard for each country to formulate its own sustainable survival strategy. Along with the chimpanzee, multitudes of other living creatures inhabit the tropics. We must implement radical reform in diplomatic policy towards government initiatives such as official development assistance (ODA), which seeks to ensure that the destruction of nature in the southern hemisphere does not advance any further. However, I fear that the government ultimately cannot solve resource problems, because each government is essentially selfish. Nor can the United Nations effectively act, as long as five countries in the Security Council hold unreasonable veto powers. Only international NGOs have any hope of finding a real solution.

The plight of the chimpanzee shows us our responsibility to pool our sagacity, define a sustainable living style, and begin living this way by all means necessary.

Postscript

One might curiously ask, 'Why on earth have you continued to conduct research for so long?' The best possible answer I could muster would be: 'Because every single thing about the chimpanzee fascinates me.' Their behaviour is so rich in variety that no matter how many years I watch them, I never grow bored.

To truly get to know their behaviour inside and out, 45 years of research is hardly enough; prolonged observation is necessary. One of the merits of long-term research is that we can acquire heaps of demographic data. It is now clear that an individual chimpanzee's life span is over 50 years, but the span of my research has just reached 45 years, thus making it difficult to come up with comprehensive answers. Much more research is required, because the population we are studying is so small: finding answers without data that cover large numbers of individuals takes years of research.

A more important benefit of long-term observation is conservation, although I did not consider this at the beginning of my research. Continuous research activity provides the indispensable information required for ecotourism. Tourists ask many questions when observing wild chimpanzees: Who is the alpha male? Who is the oldest female? What they are eating now? And so on. Only long-term research can supply the knowledge that can help ecotourism respond to the inevitable questions in an accurate and interesting way.

The constant presence of researchers makes it possible to monitor not only the health of individual chimpanzees but also changes in their environment and how these changes affect the chimpanzees.

Environmental changes are often brought about by human activities such as hunting, logging, clearing land, bush fires, and so on. After the establishment of the national park, we saw an influx of lions, African wild dogs, and spotted hyenas, even to our intensive study area, an increase in warthogs, yellow baboons and blue duikers, a drastic proliferation of alien invasive plants, particularly *Senna spectabilis*, and the virtual disappearance of trees species growing in the lower Kasiha basin, such as *Acacia albida*, *A. sieberiana*, and *Harungana madagascariensis*. Consequently, except for the invasive trees that we are making an effort to eliminate, we have seen the regeneration of natural forest and restoration of indigenous fauna and flora.

However, outside the national park, human population density has increased dramatically during the last half-century, and the periphery of the park has been largely impoverished by human exploitation. This is the time to bring a higher level of education to local residents, since education is known to be correlated all over the world to lower birth rates. The fruit of the park's revenue should also be shared more generously with the residents.

I believe the research at Mahale will continue forever, as long as human curiosity is respected.

References

Adang, O. M. J. (1984) Teasing in young chimpanzees. *Behaviour* **88**: 98–122.

Albrecht, H. & Dunnett, S. C. (1971) *Chimpanzees in Western Africa*. Piper, Munich.

Aldis, O. (1975) *Play Fighting*. Academic Press, New York.

Anderson, J. R. & McGrew, W. C. (1984) Guinea baboons (*Papio papio*) at a sleeping site. *Am J Primatol* **6**: 1–14.

Andrews, P. & Aiello, L. (1984) An evolutionary model for feeding and positional behaviour. In Chivers, D. J., Wood, B. A. & Bilsborough, A. (eds), *Food Acquisition and Processing in Primates*. Plenum Press, New York, pp. 429–466.

Azuma, S. & Toyoshima, A. (1962) Progress report of the survey of chimpanzees in their natural habitat, Kabogo Point area, Tanganyika. *Primates* **3**: 61–70.

Bateson, G. (1955) The message 'this is play'. In Schaffner, B. (ed.), *Group Processes*. Macy Foundation, New York, pp. 145–242.

Beck, B. (1980) *Animal Tool Behavior*. Garland Press, New York.

Bekoff, M. & Byers, J. A. (eds) (1998) *Animal Play*. Cambridge University Press, Cambridge.

Boehm, C. (1994) Pacifying interventions at Arnhem Zoo and Gombe. In Wrangham, R. W., McGrew, W. C., de Waal, F. B. M. & Heltne, P. G. (eds), *Chimpanzee Cultures*. Harvard University Press, Cambridge, MA, pp. 211–226.

Boehm, C. (1999) *Hierarchy in the Forest: The Evolution of Egalitarian Behavior*. Harvard University Press, Cambridge, MA.

Boesch, C. (1991a) The effects of leopard predation on grouping patterns in forest chimpanzees. *Behaviour* **117**: 220–242.

Boesch, C. (1991b) Teaching among wild chimpanzees. *Anim Behav* **41**: 530–532.

Boesch, C. (1991c) Symbolic communication in wild chimpanzees? *Hum Evol* **6**: 81–90.

Boesch, C. (1994) Cooperative hunting in wild chimpanzees. *Anim Behav* **47**: 635–667.

Boesch, C. (1995) Innovation in wild chimpanzees (*Pan troglodytes*). *Int J Primatol* **16**: 1–16.

Boesch, C. & Boesch-Achermann, H. (2000) *The Chimpanzees of the Taï Forest*. Oxford University Press, Oxford.

Boesch, C., Kohou, G., Nene, H. & Vigilant, L. (2006) Male competition and paternity in wild chimpanzees of the Taï Forest. *Am J Phys Anthropol* **130**: 103–115.

Brown, D. E. (1991) *Human Universals*. McGraw-Hill, New York.

Burghardt, G. M. (2006) *The Genesis of Animal Play*. MIT Press, Cambridge, MA.

Busse, C. (1978) Do chimpanzees hunt cooperatively? *Am Naturalist* **112**: 767–770.

Butynski, T. (2001) Africa's great apes. In Beck, B. B. (ed.) *Great Apes and Humans.* Smithsonian Institution Scholarly Press, Washington, DC, pp. 3–56.

Bygott, J. D. (1972) Cannibalism among wild chimpanzees. *Nature* **238**: 410–411.

Byrne, R. W. and Byrne, J. M. E. (1988) Leopard killers of Mahale. *Nat Hist* **97**: 22–26.

Byrne, R. W. and Byrne, J. M. E. (1993) Complex leaf-gathering skills of mountain gorillas (*Gorilla g. beringei*): variability and standardization. *Am J Primatol* **31**: 241–261.

Caldecott, J. & Miles, L. (eds) (2005) *World Atlas of Great Apes and their Conservation.* University of California Press, London.

Call, J. & Tomasello, M. (2007) The gestural repertoire of chimpanzees (*Pan troglodytes*). In Call, J. & Tomasello, M. (eds), *The Gestural Communication of Apes and Monkeys.* Lawrence Erlbaum Associates, London, pp. 17–39.

Calvert, J. J. (1985) Food selection by western gorillas (*G. g. gorilla*) in relation to food chemistry. *Oecologia* **65**: 236–246.

Caro, T. M. & Hauser, M. D. (1992) Is there teaching in non-human animals? *Quart Rev Biol* **67**: 151–174.

Carpenter, C. R. (1964) *Naturalistic Behavior of Nonhuman Primates.* Pennsylvania State University Press, University Park, PA.

Chagnon, N. A. (1992) *Yanomamo: The Last Days of Eden.* Mariner Books, San Diego.

Cheney, D. & Seyfarth, R. (2007) *Baboon Metaphysics: The Evolution of a Social Mind.* University of Chicago Press, Chicago.

Clark, C. B. (1977) A preliminary report on weaning among chimpanzees of the Gombe National Park, Tanzania. In Chevalier-Skolnikoff, S. Poirier, F. E. (eds), *Primate Bio-Social Development: Biological, Social, and Ecological Determinants.* Garland, New York, pp. 235–260.

Connor, R. C., Smolker, R. A. & Richards, A. F. (1992) Dolphin alliances and coalitions. In Harcourt, A. H. & de Waal, F. B. M. (eds), *Coalitions and Alliances in Humans and Other Animals.* Oxford University Press, Oxford, pp. 415–471.

Constable, J. L., Ashley, M. V., Goodall, J. & Pusey, A. E. (2001) Noninvasive paternity assignment in Gombe chimpanzees. *Mol Ecol* **10**: 1279–1300.

Corp, N., Hayaki, H., Matsusaka, T., *et al.* (2009) Prevalence of muzzle-rubbing and hand-rubbing behavior in wild chimpanzees in Mahale Mountains National Park, Tanzania. *Primates* **50**: 184–189.

Dawkins, R. L. (1976) *The Selfish Gene.* Oxford University Press, New York.

Diamond, J. & Bond, A. B. (1999) *Kea: Bird of Paradox.* University of California Press, Berkeley, CA.

Dixson, A. F. (1998) *Primate Sexuality.* Oxford University Press, Oxford.

Dufour, D. L. (1993) The bitter is sweet: a case study of bitter cassava (*Manihot esculenta*) use in Amazonia. In Hladik, C. M., Hladik, A., Linares, O. F., *et al.* (eds), *Tropical Forests, People and Food.* Taylor & Francis, London, pp. 575–588.

Dugatkin, L. A. (1997) *Cooperation Among Animals.* Oxford University Press, New York.

Emery-Thompson, M., Brewer-Marsden, S., Goodall, J., *et al.* (2007) Aging and fertility patterns in wild chimpanzees provide insights into the evolution of menopause. *Cur Biol* **17**: 2150–2156.

Enomoto, T. (1997) *The Bonobo.* Maruzen, Tokyo (in Japanese).

Fagen, R. (1981) *Animal Play Behavior.* Oxford University Press, Oxford.

Fawcett, K. & Muhumuza, G. (2000) Death of a wild chimpanzee community member: possible outcome of intense sexual competition? *Am J Primatol* **51**: 243–247.

Flindt, R. (2006) *Amazing Numbers in Biology*. Springer, Berlin.
Fossey, D. (1970) Making friends with mountain gorillas. *Nat Geog* **137**: 48–67.
Frisch, J. (1959) Research on primate behavior in Japan. *Amer Anthropol* **61**: 584–596.
Gallup, G. G. (1970) Chimpanzee: self-recognition. *Science* **167**: 86–87.
Gardner, R. A. & Gardner, B. T. (1969) Teaching sign language to a chimpanzee. *Science* **165**: 664–672.
Gilby, I. C. (2006) Meat sharing among the Gombe chimpanzees: harassment and reciprocal exchange. *Anim Behav* **71**: 953–963.
Goldfoot, D. A., Westerborg-van Loon, H., Groenveld, W. & Slob, A. K. (1980) Behavioral and physiological evidence of sexual climax in the female stump-tailed macaque (*Macaca arctoides*). *Science* **208**: 1477–1479.
Goldizen, A. W. (1987) Tamarins and marmosets: communal care of offspring. In Smuts, B. B., Cheney, D. L., Seyfarth, R. M., Wrangham, R. W. & Struhsaker, T. T. (eds), *Primate Societies*. University of Chicago Press, Chicago, pp. 34–43.
Goodall, J. (1963) Feeding behaviour of wild chimpanzees: a preliminary report. *Symp Zool Soc Lond* **10**: 39–48.
Goodall, J. (1968) The behaviour of free-living chimpanzees in the Gombe Stream Reserve. *Anim Behav Monogr* **1**: 161–311.
Goodall, J. (1973) Cultural elements in a chimpanzee community. In Menzel, E. W. (ed.), *Precultural Primate Behavior*. S. Karger, Basel, pp. 144–184.
Goodall, J. (1977) Infant killing and cannibalism in free-living chimpanzees. *Folia Primatol* **28**: 259–282.
Goodall, J. (1986) *The Chimpanzees of Gombe*. Harvard University Press, Cambridge, MA.
Goodall, J. (1992) Unusual violence in the overthrow of an alpha male chimpanzee at Gombe. In Nishida, T., McGrew, W. C., Marler, P., Pickford, M. & de Waal, F. B. M. (eds), *Topics in Primatology Vol. 1: Human Origins*. University of Tokyo Press, Tokyo, pp. 131–142.
Goodall, J., Bandora, A., Bergmann, E., *et al.* (1979) Inter-community interactions in the chimpanzee population of the Gombe National Park. In Hamburg, D. A. & McCown, E. R. (eds), *The Great Apes*. Benjamin/Cummings, Menlo Park, CA, pp. 13–53.
Goodman, M. (1961) The role of immunological differences in the phyletic development of human behavior. *Hum Biol* **34**: 104–150.
Groos, K. (1898) *The Play of Animals*. D. Appleton, New York.
Hamai, M. (1992) A transition of alpha-male status of wild chimpanzees at Mahale Mountains. *Primate Res* **8**: 221 (abstract).
Hamai, M., Nishida, T., Takasaki, H. & Turner, L. A. (1992) New records of within-group infanticide and cannibalism in wild chimpanzees. *Primates* **33**: 151–162.
Hamilton, W. D. (1964) The genetical evolution of social behaviour. *J Theoret Biol* **7**: 1–52.
Hamilton III, W. J. & Arrowood, P. (1978) Copulatory vocalizations of chacma baboons (*Papio ursinus*), gibbons (*Hylobates hoolock*), and humans. *Science* **200**: 1405–1409.
Hanamura, S., Kiyono, M., Nakamura, M., *et al.* (2006) A new code of observation employed at Mahale: prevention against a flu-like disease. *Pan Afr News* **13**: 1–3.
Hanamura, S., Kiyono, M., Lukasik-Braun, M., *et al.* (2008) Chimpanzee deaths at Mahale caused by a flu-like disease. *Primates* **49**: 77–80.

Harcourt, A. H. & de Waal, F. B. M. (eds) (1992) *Coalitions and Alliances in Humans and Other Animals*. Oxford University Press, Oxford.

Hasegawa, T. & Hiraiwa-Hasegawa, M. (1983) Opportunistic and restrictive matings among wild chimpanzees in the Mahale Mountains, Tanzania. *J Ethol* **1**: 75–85.

Hasegawa, T. & Hiraiwa-Hasegawa, M. (1990) Sperm competition and mating behavior. In Nishida, T. (ed.), *The Chimpanzees of the Mahale Mountains*. University of Tokyo Press, Tokyo, pp. 115–132.

Hasegawa, T., Hiraiwa-Hasegawa, M., Nishida, T. & Takasaki, H. (1983) New evidence on scavenging behavior in wild chimpanzees. *Cur Anthropol* **24**: 231–232.

Hauser, M. D. (1990) Do female chimpanzee copulation calls incite male–male competition? *Anim Behav* **39**: 596.

Hayaki, H. (1985) Social play of juvenile and adolescent chimpanzees in the Mahale Mountains National Park, Tanzania. *Primates* **26**: 343–360.

Hayaki, H., Huffman, M. A. & Nishida, T. (1989) Dominance among male chimpanzees in the Mahale Mountains National Park, Tanzania: a preliminary study. *Primates* **30**: 187–197.

Hellekant, G. & Ninomiya, Y. (1991) On the taste of *umami* in chimpanzee. *Physiol Behav* **49**: 927–934.

Hernandez-Aguilar, R. A., Moore, J. & Pickering, T. R. (2007) Savanna chimpanzees use tools to harvest the underground storage organs of plants. *Proc Natl Acad Sci USA* **104**: 19210–19213.

Hiraiwa-Hasegawa, M. (1990) Role of food sharing between mother and infant in the ontogeny of feeding behavior. In Nishida, T. (ed.), *The Chimpanzees of the Mahale Mountains*. University of Tokyo Press, Tokyo, pp. 267–275.

Hiraiwa-Hasegawa, M. (1994) Infanticide in nonhuman primates: sexual selection and local resource competition. In Parmigiani, S. & vom Saal, F. (eds), *Infanticide and Parental Care*. Harwood Academic, London, pp. 137–154.

Hiraiwa-Hasegawa, M. & Hasegawa, T. (1988) A case of offspring desertion by a female chimpanzee and the behavioral changes of the abandoned offspring. *Primates* **29**: 319–330.

Hiraiwa-Hasegawa, M., Hasegawa, M. & Nishida, T. (1984) Demographic study of a large-sized unit-group of chimpanzees in the Mahale Mountains, Tanzania: a preliminary report. *Primates* **25**: 401–413.

Hiraiwa-Hasegawa, M., Byrne, R. W., Takasaki, H. &, Byrne J. E. (1986) Aggression toward large carnivores by wild chimpanzees of Mahale Mountains National Park, Tanzania. *Folia Primatol* **47**: 8–13.

Hladik, C. M. (1981) Diet and the evolution of feeding strategies among forest primates. In Harding, R. S. O. & Teleki, G. (eds), *Omnivorous Primates*. Columbia University Press, New York, pp. 215–254.

Hladik, C. M. (1993) Fruits of the rain forest and taste perception as a result of evolutionary interactions. In Hladik, C. M., Hladik, A., Linares, O. F., *et al.* (eds), *Tropical Forests, People and Food*. Taylor & Francis, London, pp. 73–82.

Hladik, C. M. & Simmen, B. (1996) Taste perception and feeding behavior in nonhuman primates and human populations. *Evol Anthropol* **5**: 58–71.

Horai, S., Satta, Y., Hayasaka, K., *et al.* (1992) Man's place in Hominoidea revealed by mitochondrial DNA genealogy. *J Mol Evol* **35**: 32–43.

Hosaka, K. (1995a) A single flu epidemic killed at least 11 chimpanzees. *Pan Afr News* **2**(2): 3–4.

Hosaka, K. (1995b) A rival yesterday is a friend today: a grand political drama in the forest. *Pan Afr News* **2**(2): 10–11.

Hosaka, K. & Nishida, T. (2002) Restoration of an alpha male from ostracism. In Nishida, T., Uehara, S. & Kawanaka, K. (eds), *The Mahale Chimpanzees: Thirty-seven Years of Panthropology*. Kyoto University Press, Kyoto, pp. 439–471 (in Japanese).

Hosaka, K., Nishida, T., Hamai, M., Matsumoto-Oda, A. & Uehara, S. (2001) Predation of mammals by the chimpanzees of the Mahale Mountains, Tanzania. In Galdikas, B. M. F., Briggs, N. E., Sheeran, L. K., Shapiro, G. L. & Goodall, J. (eds), *All Apes, Great and Small, Vol.1: African Apes*. Kluwer Academic, New York, pp. 107–130.

Hosaka, K., Inoue, E. & Fujimoto, M. (2008) Reactions of wild chimpanzees to aardvark (*Orycteropus afer*) carcasses in the Mahale Mountains National Park. In Nishida, T. (ed.), *Emergence of Novel Behavior Pattern and Developmental Process of Cultural Behavior: Report to the Ministry of Education, Culture, Sports, Science and Technology*, Book Print Center. Kyoto Business Association of University Cooperatives, Kyoto, pp. 34–39.

Hrdy, S. B. (1979) Infanticide among animals: a review, classification, and examination of the implications for the reproductive strategies of females. *Ethol & Sociobiol* **1**: 13–40.

Huffman, M. A. (1997) Current evidence for self-medication in primates: a multidisciplinary perspective. *Yb Phys Anthropol* **40**: 171–200.

Huffman, M. A. & Kalunde, M. S. (1993) Tool-assisted predation of a squirrel by a female chimpanzee in the Mahale Mountains, Tanzania. *Primates* **34**: 93–98.

Huffman, M. A. & Seifu, M. (1989) Observations on the illness and consumption of a possibly medicinal plant *Vernonia amygdalina* (Del), by a wild chimpanzee in the Mahale Mountains National Park, Tanzania. *Primates* **30**: 51–63.

Huizinga, J. (1950) *Homo ludens*. Beacon, Boston.

Idani, G. (1991) Social relationships between immigrant and resident bonobo (*Pan paniscus*) females at Wamba. *Folia Primatol* **57**: 83–95.

Idani, G. (1995) Function of peering behavior among bonobos (*Pan paniscus*) at Wamba. *Primates* **36**: 377–383.

Ihobe, H. (2002) Anti-chimpanzee strategies of red colobus monkeys. In Nishida, T., Uehara, S. & Kawanaka, K. (eds), *The Mahale Chimpanzees: Thirty-seven Years of Panthropology*. Kyoto University Press, Kyoto, pp. 245–260 (in Japanese).

Imanishi, K. (1958a) Gorillas: a preliminary survey in 1958. *Primates* **1**: 73–78.

Imanishi, K. (1958b) The origin of human family: a primatological approach. *Jpn J Ethnol* **25**: 110–130.

Imanishi, K. (2002 [1941]) *The World of Living Things*. Routledge Curzon, London.

Inaba, A. (2009) Power takeover occurred in M group of the Mahale Mountains, Tanzania, in 2007. *Pan Afr News* **16**: 13–15.

Inagaki, H. & Tsukahara, T. (1993) A method of identifying chimpanzee hairs in lion feces. *Primates* **34**: 109–112.

Inoue, E., Inoue-Murayama, M., Vigilant, L., Takenaka, O. & Nishida, T. (2008) Relatedness in wild chimpanzees: influence of paternity, male philopatry, and demographic factors. *Am J Phys Anthropol* **137**: 256–262.

Isaac, G. (1978) Food sharing behavior of proto-human hominids. *Scient Amer* **238**: 90–109.

Itani, J. (1979) Distribution and adaptation of chimpanzees in an arid area. In Hamburg, D. A. & McCown, E. R. (eds), *The Great Apes*. Benjamin/Cummings, Menlo Park, CA, pp. 55–72.

Itani, J. & Suzuki, A. (1967) The social unit of chimpanzees. *Primates* **8**: 355–381.

Itoh, N. (2002) Food in the forest: density and spatial distribution of chimpanzee foods. In Nishida, T., Uehara, S. & Kawanaka, K. (eds), *Mahale Chimpanzees: Thirty-seven Years of Panthropology*. Kyoto University Press, Kyoto, pp. 77–100 (in Japanese).

Itoh, N. & Nishida, T. (2007) Chimpanzee grouping patterns and food availability in Mahale Mountains National Park, Tanzania. *Primates* **48**: 87–96.

Izawa, K. (1970) Unit groups of chimpanzees and their nomadism in the savanna woodland. *Primates* **11**: 1–46.

Izawa, K. (1975) *Good Bye Brucy*. NHK Publishing Association, Tokyo (in Japanese).

Janzen, D. H. (1983) Dispersal of seeds by vertebrate guts. In Futuyma, D. J. & Slatkin, M. (eds), *Coevolution*. Sinauer, Sunderland, pp. 232–262.

Jisaka, M., Ohigashi, H., Takegawa, K., Huffman, M. A. &, Koshimizu, K. (1993) Antitumor and antimicrobial activities of bitter sesquiterpene lactones of *Vernonia amygdalina*, a possible medicinal plant used by wild chimpanzees. *Biosci Biotech Biochem* **57**: 833–834.

Johns, T. (1989) A chemical–ecological model of root and tuber domestication in the Andes. In Harris, D. R. & Hillman, G. C. (eds), *Foraging and Farming: The Evolution of Plant Exploitation*. Unwin Hyman, London, pp. 504–519.

Kahlenberg, S., Emery Thompson, M. & Wrangham, R. W. (2008) Female competition over core areas among *Pan troglodytes schweinfurthii*, Kibale National Park, Uganda. *Int J Primatol* **29**: 931–948.

Kalmus, H. (1970) The sense of taste of chimpanzees and other primates. In Bourne, G. H. (ed.), *The Chimpanzee, Vol. 2*. Karger Basel, pp. 130–141.

Kano, T. (1972) Distribution and adaptation of the chimpanzee in the open country on the eastern shore of Lake Tanganyika. *Kyoto Univ Afr Studies* **7**: 37–129.

Kano, T. (1979) A pilot study on the ecology of pygmy chimpanzees (*Pan paniscus*). In Hamburg, D. A. & McCown, E. R. (eds), *The Great Apes*. Benjamin/Cummings, Menlo Park, CA, pp. 123–135.

Kano, T. (1980) Social behavior of wild pygmy chimpanzees (*Pan paniscus*) of Wamba: a preliminary report. *J Hum Evol* **9**: 243–260.

Kano, T. (1984) Observations of physical abnormalities among the wild bonobos (*Pan paniscus*) of Wamba, Zaire. *Am J Phys Anthropol* **63**: 1–11.

Kano, T. (1992) *The Last Ape*. Stanford University Press, Stanford, CA.

Kano, T. (1998) A preliminary glossary of bonobo behaviors at Wamba. In Nishida, T. (ed.), *Comparative Study of the Behavior of the Genus Pan by Compiling Video Ethogram: Report to the Ministry of Education, Culture, Sports, Science and Technology*. Nissho Printer, Kyoto, pp. 39–81.

Katayose, T. (1963) *The Song of Bwana Toshi*. Adahi-Shinbun-sha, Tokyo (in Japanese).

Kaur, T., Singh, J., Tong, S., *et al.* (2008) Descriptive epidemiology of fatal respiratory outbreaks and detection of a human-related metapneumovirus in wild chimpanzees (*Pan troglodytes*) at Mahale Mountains National Park, western Tanzania. *Am J Primatol* **70**: 755–765.

Kawabata, M. & Nishida, T. (1991) A preliminary note on the intestinal parasites of wild chimpanzees in the Mahale Mountains, Tanzania. *Primates* **32**: 275–278.

Kawai, M. (1958) On the system of social ranks in a natural group of Japanese monkeys. *Primates* **1**: 11–48.

Kawanaka, K. (1990) Alpha males' interactions and social skills. In Nishida, T. (ed.), *The Chimpanzees of the Mahale Mountains*. University of Tokyo Press, Tokyo, pp. 171–187.

Kawanaka, K. (2002) Changes in the social relationships among high-ranking adult males. In Nishida, T., Uehara, S. & Kawanaka, K. (eds), *The Mahale Chimpanzees: Thirty-seven Years of Panthropology*. Kyoto University Press, Kyoto, pp. 417–438 (in Japanese).

Kokwaro, J. O. (1976) *Medicinal Plants of East Africa*. East African Literature Bureau, Nairobi.

Köndgen, S., Köhl, H., N'Goran, P. K., *et al.* (2008) Pandemic human viruses cause decline of endangered great apes. *Curr Biol* **18**: 260–264.

Kormos, R., Boesch, C., Bakarr, M. I. & Butynski, T. M. (eds) (2003) *West African Chimpanzees: Status Survey and Conservation Action Plan*. IUCN, Gland, Switzerland.

Kortlandt, A. (1962) Chimpanzees in the wild. *Scient Amer* **206**: 128–138.

Kortlandt, A. (1967) Experimentation with chimpanzees in the wild. In Starck, D., Schneider, R. & Kuhn, H.-J. (eds), *Neue Ergebnisse der Primatologia*. Fischer, Stuttgart, pp. 208–224.

Kortlandt, A. & Kooij, M. (1963) Protohominid behaviour in primates. *Symp Zool Soc Lond* **10**: 61–87.

Koshimizu, K., Ohigashi, H., Huffman, M. A., Nishida, T. & Takasaki, H. (1993) Physiological activities and the active constituents of potentially medicinal plants used by wild chimpanzees of the Mahale Mountains, Tanzania. *Int J Primatol* **14**: 345–356.

Kummer, H. (1971) *Primate Societies*. Aldine, Chicago.

Kummer, H. (1995) *In Quest of the Sacred Baboon*. Princeton University Press, Princeton, NJ.

Kuroda, S. (1979) Grouping of the pygmy chimpanzees. *Primates* **20**: 161–183.

Kuroda, S. (1980) Social behavior of pygmy chimpanzees. *Primates* **21**: 181–197.

Lancaster, J. B. (1972) Play-mothering: the relations between juvenile females and young infants among free-ranging vervets. In Poirier, F. E. (ed.), *Primate Socialization*. Random House, New York.

Langergraber, K. E., Mitani, J. C. & Vigilant, L. (2007) The limited impact of kinship on cooperation in wild chimpanzees. *Proc Natl Acad Sci USA* **104**: 7786–7790.

Lee, R. B. (1979) *The !Kung San*. Cambridge University Press, Cambridge.

Lehmann, J. & Boesch, C. (2004) To fission or to fusion: effects of community size on wild chimpanzee (*Pan troglodytes verus*) social organisation. *Behav Ecol Sociobiol* **56**: 207–216.

Leopold, A. C. & Ardrey, R. (1972) Toxic substances in plants and the food habits of early man. *Science* **176**: 512–513.

Lorenz, K. (1966) *On Aggression*. Harcourt, Brace and World, New York.

Lukosi, N. (1997) A brief note on possible control of *Senna spectabilis*, an invasive exotic tree at Mahale. *Pan Afr News* **4**: 18.

Matsumoto-Oda, A., Hosaka, K., Huffman, M. A. & Kawanaka, K. (1998) Factors affecting party size in chimpanzees of the Mahale Mountains. *Int J Primatol* **19**: 999–1011.

Matsusaka, T. (2004) When does play panting occur during social play in wild chimpanzees? *Primates* **45**: 221–229.

Matsusaka, T., Nishie, H., Shimada, M., *et al.* (2006) Tool-use for drinking water by immature chimpanzees of Mahale: prevalence of an unessential behavior. *Primates* **47**: 113–122.

McGinnis, P. A. (1979) Sexual behavior in free-living chimpanzees: consort relationships. In Hamburg, D. A. & McCown, E. R. (eds), *The Great Apes*. Benjamin/Cummings, Menlo Park, CA, pp. 429–439.

McGrew, W. C. (1974) Tool use by wild chimpanzees in feeding upon driver ants. *J Hum Evol* **3**: 501–508.

McGrew, W. C. (1979) Evolutionary implications of sex differences in chimpanzee predation and tool use. In Hamburg, D. A. & McCown, E. R. (eds), *The Great Apes*. Benjamin/Cummings, Menlo Park, CA, pp. 440–463.

McGrew, W. C. (1983) Animal foods in the diet of wild chimpanzees (*Pan troglodytes*): why cross-cultural variation? *J Ethol* **1**: 46–61.
McGrew, W. C. (1992) *Chimpanzee Material Culture: Implications for Human Evolution.* Cambridge University Press, Cambridge.
McGrew, W. C. & Collins, D. A. (1985) Tool use by wild chimpanzees (*Pan troglodytes*) to obtain termites (*Macrotermes herus*) in the Mahale Mountains, Tanzania. *Am J Primatol* **9**: 47–62.
McGrew, W. C. & Tutin, C. E. G. (1978) Evidence for a social custom in wild chimpanzees? *Man* **13**: 234–251.
McGrew, W. C., Baldwin, P. J. & Tutin, C. E. G. (1981) Chimpanzees in a hot, dry and open habitat: Mt. Assirik, Senegal, West Africa. *J Hum Evol* **10**: 227–244.
McGrew, W. C., Marchant, F. L., Scott, S. E. & Tutin, C. E. G. (2001) Intergroup differences in a social custom of wild chimpanzees: the grooming hand-clasp of the Mahale Mountains. *Cur Anthropol* **42**: 148–153.
Michael, R. O. & Keverne, E. B. (1970) Primate sex pheromones of vaginal origin. *Nature* **225**: 84–85.
Mitani, J. C. & Nishida, T. (1993) Contexts and social correlates of long-distance calling by male chimpanzees. *Anim Behav* **45**: 735–746.
Mitani, J. C. & Watts, D. P. (1999) Demographic influence on the hunting behavior of chimpanzees. *Am J Phys Anthropol* **109**: 439–454.
Mitani, J. C. & Watts, D. P. (2001) Why do chimpanzees hunt and share meat? *Anim Behav* **61**: 915–924.
Mitani, J. C., Hasegawa, T., Gros-Louis, J. & Marler, P. (1992) Dialects in wild chimpanzees? *Am J Primatol* **27**: 233–243.
Mitani, J. C., Watts, D. P. & Muller, M. N. (2002) Recent developments in the study of wild chimpanzee behavior. *Evol Anthropol* **11**: 9–25.
Mitani, J. C., Watts, D. P. & Amsler, S. J. (2010) Lethal intergroup aggression leads to territorial expansion in wild chimpanzees. *Cur Biol* **20**(12): 1–2.
Morris, D. (1981) *The Soccer Tribe.* Jonathan Cape, London.
Moss, C. (1989). *Elephant Memories: Thirteen Years in the Life of an Elephant Family.* Fontana/Collins, Glasgow.
Muller, M. N. (2002) Agonistic relations among Kanyawara chimpanzees. In Boesch, C., Hohmann, G. & Marchant, L. F. (eds), *Behavioural Diversity in Chimpanzees and Bonobos.* Cambridge University Press, Cambridge, pp. 112–124.
Murata, G. & Hazama, N. (1968) *The Food Plants of Japanese Macaques in Arashiyama.* Iwatayama Natural History Institute, Kyoto.
Nadler, R. D. (1976) Sexual behavior of captive lowland gorillas. *Arch Sex Behav* **5**: 487–502.
Nadler, R. D. (1995) Sexual behavior of orangutans (*Pongo pygmaeus*). In Nadler, R. D., Galdikas, B. F. M., Sheeran, L. K. & Rosen, N. (eds), *The Neglected Ape.* Plenum, New York, pp. 223–237.
Nakamura, M. (2002) Grooming-hand-clasp in Mahale M group chimpanzees: implications for culture in social behaviours. In Boesch, C., Hohmann, G. & Marchant, L. F. (eds), *Behavioural Diversity in Chimpanzees and Bonobos.* Cambridge University Press, Cambridge, pp. 71–83.
Nakamura, M. (2003) 'Gatherings' of social grooming among wild chimpanzees: implications for evolution of sociality. *J Hum Evol* **44**: 59–71.
Nakamura, M. (2010) Ubiquity of culture and possible social inheritance of sociality among wild chimpanzees. In Lonsdorf, E. V., Ross, S. R. & Matsuzawa, T. (eds), *The Mind of the Chimpanzee.* University of Chicago Press, Chicago, pp. 156–167.
Nakamura, M. & Itoh, N. (2005) Notes on the behavior of a newly immigrated female chimpanzee to the Mahale M group. *Pan Afr News* **12**: 20–22.

Nakamura, M. & Itoh, N. (2008) Hunting with tools by Mahale chimpanzees. *Pan Afr News* **15**: 3–6.
Nakamura, M. & Nishida, T. (2008) Developmental process of grooming-hand-clasp by chimpanzees of the Mahale Mountains, Tanzania. *Primate Eye* **96**: 247.
Nakamura, M. & Nishida, T. (2009) Chimpanzee tourism in relation to the viewing regulations at the Mahale Mountains National Park, Tanzania. *Primate Conserv* **24**: 1–6.
Nakamura, M. & Uehara, S. (2004) Proximate factors of different types of grooming-hand-clasp in Mahale chimpanzees: implications for chimpanzee social customs. *Cur Anthropol* **45**: 108–114.
Nakamura, M., McGrew, W. C., Marchant, L. F. & Nishida, T. (2000) Social scratch: another custom in wild chimpanzees? *Primates* **41**: 237–248.
Napiers, J. R. & Napier, P. H. (1967) *A Handbook of Living Primates*. Academic Press, London.
Newton, P. & Nishida, T. (1991) Possible buccal administration of herbal drugs by wild chimpanzees (*Pan troglodytes*). *Anim Behav* **39**: 799–800.
Nishida, T. (1968) The social group of wild chimpanzees in the Mahali Mountains. *Primates* **9**: 167–224.
Nishida, T. (1970) Social behavior and relationship among wild chimpanzees of the Mahali Mountains. *Primates* **11**: 47–87.
Nishida, T. (1972a) Preliminary information of the pygmy chimpanzees of the Congo Basin. *Primates* **13**: 415–425.
Nishida, T. (1972b) A note on the ecology of the red colobus monkeys (*Colobus badius tephrosceles*) living in the Mahali Mountains. *Primates* **3**: 57–64.
Nishida, T. (1972c) Tool use by wild chimpanzees. *Shizen* **27**(8): 41–47. (In Japanese)
Nishida, T. (1973a) *The Children of the Mountain Spirits*. Chikuma-shobo, Tokyo (in Japanese).
Nishida, T. (1973b) The ant-gathering behaviour by the use of tools among wild chimpanzees of the Mahali Mountains. *J Hum Evol* **2**: 357–370.
Nishida, T. (1974) Ecology of wild chimpanzees. In Ohtsuka, R., Tanaka, J. & Nishida, T. (eds), *Human Ecology*. Kyoritsu-Shuppan, Tokyo, pp. 15–60 (in Japanese).
Nishida, T. (1976) The bark-eating habits in primates, with special reference to their status in the diet of wild chimpanzees. *Folia Primatol* **25**: 277–287.
Nishida, T. (1979) The social structure of chimpanzees of the Mahale Mountains. In Hamburg, D. A. & McCown, E. R. (eds), *The Great Apes*. Benjamin/Cummings, Menlo Park, CA, pp. 73–121.
Nishida, T. (1980) The leaf-clipping display: a newly-discovered expressive gesture in wild chimpanzees. *J Hum Evol* **9**: 117–128.
Nishida, T. (1981) *The World of Wild Chimpanzees*. Chuokoron-sha, Tokyo (in Japanese).
Nishida, T. (1983a) Alpha status and agonistic alliance in wild chimpanzees (*Pan troglodytes schweinfurthii*). *Primates* **24**: 318–336.
Nishida, T. (1983b) Alloparental behavior in wild chimpanzees of the Mahale Mountains, Tanzania. *Folia Primatol* **41**: 1–33.
Nishida, T. (1987) Local traditions and cultural transmission. In Smuts, B. B., Cheney, D. L., Seyfarth, R. M., Wrangham, R. W. & Struhsaker, T. T. (eds), *Primate Societies*. University of Chicago Press, Chicago, pp. 462–474.
Nishida, T. (1988) Development of social grooming between mother and offspring in wild chimpanzees. *Folia Primatol* **50**: 109–123.

Nishida, T. (1989) Social interactions between resident and immigrant female chimpanzees. In Heltne, P. G. & Marquardt, L. A. (eds), *Understanding Chimpanzees*. Harvard University Press, Cambridge, MA, pp. 68–89.
Nishida, T. (ed.) (1990a) *The Chimpanzees of the Mahale Mountains*. University of Tokyo Press, Tokyo.
Nishida, T. (1990b) A quarter century of research in the Mahale Mountains: an overview. In Nishida, T. (ed.), *The Chimpanzees of the Mahale Mountains*. University of Tokyo Press, Tokyo, pp. 3–35.
Nishida, T. (1990c) Deceptive behavior in young chimpanzees: an essay. In Nishida, T. (ed.), *The Chimpanzees of the Mahale Mountains*. University of Tokyo Press, Tokyo, pp. 285–290.
Nishida, T. (1991a) Weaning conflict in chimpanzees. Lecture to Second International Symposium on 'Understanding Chimpanzees', Chicago Academy of Sciences, Chicago, December 1991.
Nishida, T. (1991b) Primate gastronomy. In Friedman, M. I., Tordoff, M. G. & Kare, M. R. (eds), *Appetite and Nutrition*. Marcell Dekker, New York, pp. 195–209.
Nishida, T. (1992) The evolution of aid-giving behavior in primates. In Shibatani, A., Nagano, K. & Yourou, T. (eds), *Evolution, Vol.7: Evolution from Ecological Perspectives*. University of Tokyo Press, Tokyo, pp. 247–306 (in Japanese).
Nishida, T. (1993a) Left nipple suckling preference in wild chimpanzees. *Ethol Sociobiol* **14**: 45–52.
Nishida, T. (1993b) Chimpanzees are always new to me. In Cavalieri, P. & Singer, P. (eds), *The Great Ape Project*. Fourth Estate, London, pp. 24–26.
Nishida, T. (1994) *Thirty-six Stories of Wild Chimpanzees*. Kinokuniya-shoten, Tokyo (in Japanese).
Nishida, T. (1996a) The death of Ntologi: the unparalleled leader of M group. *Pan Afr News* **3**(1): 4.
Nishida, T. (1996b) Eradication of the invasive, exotic tree *Senna spectabilis* in the Mahale Mountains. *Pan Afr News* **3**(2): 6–7.
Nishida, T. (1997a) Sexual behavior of adult male chimpanzees of the Mahale Mountains National Park, Tanzania. *Primates* **38**: 379–398.
Nishida, T. (1997b) Baboon invasion into chimpanzee habitat. *Pan Afr News* **4**: 11–12.
Nishida, T. (1998) Deceptive tactic by an adult male chimpanzee to snatch a dead infant from its mother. *Pan Afr News* **5**: 13–15.
Nishida, T. (1999) *Where Does Human Nature Come From?* Kyoto University Press, Kyoto. (In Japanese)
Nishida, T. (2001) Notes on incest taboo. In Kawada, J. (ed.), *Incest and its Taboo*. Fujiwara-shoten, Tokyo, pp. 137–145 (in Japanese).
Nishida, T. (2003a) Harassment of mature female chimpanzees by young males in the Mahale Mountains. *Int J Primatol* **24**: 503–514.
Nishida, T. (2003b) Individuality and flexibility of cultural behavior patterns in chimpanzees. In de Waal, F. B. M. & Tyack, P. L. (eds), *Animal Social Complexity*. Harvard University Press, Cambridge, MA, pp. 392–413.
Nishida, T. (2004) Lack of 'group play' in wild chimpanzees. *Pan Afr News* **11**: 2–3.
Nishida, T. (2008a) Why were guava trees cut down in Mahale Park? The question of exterminating all introduced plants. *Pan Afr News* **15**: 12–14.
Nishida, T. (2008b) *The Society of Chimpanzees*. Toho-shuppan, Osaka. (In Japanese)
Nishida, T. (2009) The 60th anniversary of Japanese primatology. *Primates* **50**: 1–2.

Nishida, T. & Hiraiwa, M. (1982) Natural history of a tool-using behavior by wild chimpanzees in feeding upon wood-boring ants. *J Hum Evol* **11**: 73–99.

Nishida, T. & Hiraiwa-Hasegawa, M. (1985) Responses to a stranger mother–son pair in the wild chimpanzee: a case report. *Primates* **26**: 1–13.

Nishida, T. & Hiraiwa-Hasegawa, M. (1987) Chimpanzees and bonobos: cooperative relationships among males. In Smuts, B. B., Cheney, D. L., Seyfarth, R. M., Wrangham, R. W. & Struhsaker, T. T. (eds), *Primate Societies*. University of Chicago Press, Chicago, pp. 165–177.

Nishida, T. & Hosaka, K. (1996) Coalition strategies among adult male chimpanzees of the Mahale Mountains. In McGrew, W. C., Marchant, L. F. & Nishida, T. (eds), *Great Ape Societies*. Cambridge University Press, Cambridge, pp. 114–134.

Nishida, T. & Inaba, A. (2009) Pirouettes: the rotational play of wild chimpanzees. *Primates* **50**: 333–341.

Nishida, T. & Kawanaka, K. (1972) Inter-unit-group relationships among wild chimpanzees of the Mahali Mountains. *Kyoto Univ Afr Stud* **7**: 131–169.

Nishida, T. & Kawanaka, K. (1985) Within-group cannibalism by adult male chimpanzees. *Primates* **26**: 274–284.

Nishida, T. & Nakamura, M. (1993) Chimpanzee tool-use to clear a blocked nasal passage. *Folia Primatol* **61**: 218–220.

Nishida, T. & Nakamura, M. (2008) Long-term research and conservation in the Mahale Mountains, Tanzania. In Wrangham, R. W. & Ross, E. (eds), *Science and Conservation in African Forests: The Benefits of Long-term Research*. Cambridge University Press, Cambridge, pp. 173–183.

Nishida, T. & Turner, L. A. (1996) Food transfer between mother and infant chimpanzees of the Mahale Mountains National Park, Tanzania. *Int J Primatol* **17**: 947–968.

Nishida, T. & Uehara, S. (1980) Chimpanzees, tools, and termites: another example from Tanzania. *Cur Anthropol* **21**: 671–672.

Nishida, T. & Uehara, S. (1981) Kitongwe names of plants: a preliminary listing. *Afr Stud Monogr* **1**: 109–131.

Nishida, T. & Uehara, S. (1983) Natural diet of chimpanzees (*Pan troglodytes schweinfurthii*): long-term record from the Mahale Mountains, Tanzania. *Afr Stud Monogr* **3**: 109–130.

Nishida, T. & Wallauer, W. (2003) Leaf-pile pulling: an unusual play pattern in wild chimpanzees. *Am J Primatol* **60**: 167–173.

Nishida, T. & Zamma, K. (2000) Responses of chimpanzees to a fresh carcass of a leopard. Abstract, 37th Congress of Japanese Association of African Studies, p. 56.

Nishida, T., Uehara, S. & Nyundo, R. (1979) Predatory behavior among wild chimpanzees of the Mahale Mountains. *Primates* **20**: 1–20.

Nishida, T., Wrangham, R. W., Goodall, J. & Uehara, S. (1983) Local differences in plant-feeding habits of chimpanzees between the Mahale Mountains and Gombe National Park, Tanzania. *J Hum Evol* **12**: 467–480.

Nishida, T., Hiraiwa-Hasegawa, M., Hasegawa, T. & Takahata, Y. (1985) Group extinction and female transfer in wild chimpanzees in the Mahale National Park, Tanzania. *Z Tierpsychol* **67**: 284–301.

Nishida, T., Takasaki, H. & Takahata, Y. (1990) Demography and reproductive profiles. In Nishida, T. (ed.), *The Chimpanzees of the Mahale Mountains*. University of Tokyo Press, Tokyo, pp. 63–97.

Nishida, T., Hasegawa, T., Hayaki, H., Takahata, Y. & Uehara, S. (1992) Meat-sharing as a coalition strategy by an alpha male chimpanzee? In Nishida, T.,

McGrew, W.C., Marler, P., Pickford, M. & de Waal, F.B.M. (eds), *Topics in Primatology, Vol. 1: Human Origins*. University of Tokyo Press, Tokyo, pp. 159–174.
Nishida, T., Hosaka, K., Nakamura, M. & Hamai, M. (1995) A within-group gang attack on a young adult male chimpanzee: ostracism of an ill-mannered member? *Primates* **36**: 207–211.
Nishida, T., Kano, T., Goodall, J., McGrew, W.C. & Nakamura, M. (1999) Ethogram and ethnography of Mahale chimpanzees. *Anthropol Sci* **107**: 141–188.
Nishida, T., Ohigashi, H. & Koshimizu, K. (2000) Taste of chimpanzee plant foods. *Cur Anthropol* **41**: 431–438.
Nishida, T., Corp, N., Hamai, M., Hasegawa, T., Hiraiwa-Hasegawa, M., Hosaka, K., *et al.* (2003) Demography, female life history and reproductive profiles among the chimpanzees of Mahale. *Am J Primatol* **59**: 99–121.
Nishida, T., Matsusaka, T. & McGrew, W.C. (2009) Emergence, propagation or disappearance of novel behavioral patterns in the habituated chimpanzees of Mahale: a review. *Primates* **50**: 23–36.
Nishida, T., Zamma, K., Matsusaka, T., Inaba, A. & McGrew, W.C. (2010) *Chimpanzee Behavior in the Wild: An Audio-Visual Encyclopedia*. Springer, Tokyo.
Nishie, H. (2004a) Disappearance of Fanana and subsequent incidents. *Mahale Chimpun* **3**: 5–6 (in Japanese).
Nishie, H. (2004b) Fall of Fanana and enthronement of Alofu. *Mahale Chimpun* **4**: 6–7 (in Japanese).
Nissen, H. (1931) A field study of the chimpanzee. *Comp Psychol Monogr* **8**: 1–121.
Ogawa, H., Kanamaori, M. & Mukeni, H. (1997) The discovery of chimpanzees in the Lwazi River area, Tanzania: a new southern distribution limit. *Pan Afr News* **4**: 1–3.
Ohashi, G. (2007) Papaya fruit sharing in wild chimpanzees at Bossou, Guinea. *Pan Afr News* **14**: 14–16.
Ohigashi, H., Huffman, M.A., Izutsu, D., *et al.* (1994) Toward the chemical ecology of medicinal plant-use in chimpanzees: the case of *Vernonia amygdalina* (Del). A plant used by wild chimpanzees possibly for parasite-related diseases. *J Chem Ecol* **20**: 541–553.
Page, J.E., Balza, F., Nishida, T. & Towers, G.H.N. (1992) Biologically active diterpenes from *Aspilia mossambicensis*, a chimpanzee medicinal plant. *Phytochem* **31**: 3437–3439.
Patterson, J.H. (1979) *The Man-eaters of Tsavo*. Macmillan, London.
Pellegrini, A.D. & Smith, P.K. (eds) (2005) *The Nature of Play*. Guilford Press, New York.
Perry, S. (2008) *Manipulative Monkeys: The Capuchins of Lomas Barbudal*. Harvard University Press, Cambridge, MA.
Pika, S. (2007) Gestures in subadult bonobos. In Call, J. & Tomasello, M. (eds), *The Gestural Communication of Apes and Monkeys*. Laurence Erlbaum Associates, London, pp. 41–67.
Power, T.G. (2000) *Play and Exploration in Children and Animals*. Lawrence Erlbaum Associates, Mahwah, NJ.
Pruetz, J.D., & Bertolani, P. (2007) Savanna chimpanzees, *Pan troglodytes verus*, hunt with tools. *Cur Biol* **17**: 412–417.
Pusey, A.E. (1990) Behavioural changes at adolescence in chimpanzees. *Behaviour* **115**: 203–246.
Pusey, A.E., Oehlet, C.W., Williams, J. & Goodall, J. (2005) Influence of ecological and social factors on body mass of wild chimpanzees. *Int J Primatol* **26**: 3–31.

Pusey, A. E., Murray, C., Wallauer, W., *et al.* (2008) Severe aggression among female *Pan troglodytes schweinfurthii* at Gombe National Park, Tanzania. *Int J Primatol* **29**: 949–974.

Reynolds, V. (1966) Open groups in hominid evolution. *Man* **1**: 441–452.

Reynolds, V. & Reynolds, F. (1965) Chimpanzees of the Budongo Forest. In DeVore, I. (ed.), *Primate Behavior*. Holt, Rinehart & Winston, New York, pp. 368–424.

Rodriguez, E., Aregullin, M., Nishida, T., *et al.* (1985) Thiarubrine-A, a bioactive constituent of *Aspilia* (Asteraceae) consumed by wild chimpanzees. *Experientia* **41**: 419–420.

Rogers, M. E., Maisels, F., Williamson, E. A., Fernandez, M. & Tutin, C. E. G. (1990) Gorilla diet in the Lope Reserve, Gabon. *Oecologia* **84**: 326–339.

Sabater Pi, J. (1977) Contribution to the study of alimentation of lowland gorillas in the natural state, in Rio Muni (West Africa). *Primates* **18**: 183–204.

Sabater Pi, J. (1979) Feeding behaviour and diet of chimpanzees (*Pan troglodytes troglodytes*) in the Okorobiko Mountains in Rio Muni (West Africa). *Z Tierpsychol* **50**: 165–281.

Saitoh, K. (1989) *Imanishi Kinji: Seeking for Nature*. Shoraisha, Kyoto (in Japanese).

Sakamaki, T. (1998) First record of algae-feeding by a female chimpanzee at Mahale. *Pan Afr News* **5**: 1–3.

Sakamaki, T., Itoh, N. & Nishida, T. (2001) An attempted within-group infanticide in wild chimpanzees. *Primates* **42**: 359–366.

Sakura, O. (1994) Factors affecting party size and composition of chimpanzees (*Pan troglodytes verus*) at Bossou, Guinea. *Int J Primatol* **15**: 167–183.

Sanz, C., Morgan, D. & Gulick, S. (2004) New insights into chimpanzees, tools, and termites from the Congo Basin. *Am Nat* **164**: 567–581.

van Schaik, C. P. (2000) Social counterstrategies against infanticide by males in primates and other mammals. In Kappeler, P. (ed.), *Primate Males*. Cambridge University Press, Cambridge, pp. 34–63.

Schaller, G. B. (1963) *The Mountain Gorilla: Ecology and Behavior*. University of Chicago Press, Chicago.

Schick, K. D. & Toth, N. (1993) *Making Silent Stones Speak*. Simon & Schuster, New York.

Shimada, M. (2002) Social scratch among chimpanzees in Gombe. *Pan Afr News* **9**: 21–23.

Short, R. V. (1979) Sexual selection and its component parts, somatic and genital selection as illustrated by man and the great apes. *Adv Stud Behav* **9**: 131–158.

Silberbauer, G. (1981) Hunter/gatherers of the central Kalahari. In Harding, R. S. O. & Teleki, G. (eds), *Omnivorous Primates*. Columbia University Press, New York, pp. 455–498.

Silk, J. (1978) Patterns of food sharing among mother and infant chimpanzees at Gombe National Park, Tanzania. *Folia Primatol* **29**: 129–141.

Stahl, A. B. (1984) Hominid dietary selection before fire. *Cur Anthropol* **25**: 151–168.

Stanford, C. B. (1998) *Chimpanzee and Red Colobus*. Harvard University Press, Cambridge, MA.

Stanley, H. M. (1878) *Through the Dark Continent*. Sampson Low, Marston, London.

Strier, K. (1998) Menu for a monkey. In Ciochon, R. L. & Nisbett, R. A. (eds), *The Primate Anthology*. Prentice Hall, Upper Saddle River, NJ, pp. 180–186.

Struhsaker, T. T. (1975) *The Red Colobus Monkey*. University of Chicago Press, Chicago.

Stumpf, R. (2007) Chimpanzees and bonobos: diversity within and between species. In Campbell, C. J., Fuentes, A., MacKinnon, K. C., Panger, M. &

Bearder, S. K. (eds), *Primates in Perspective*. Oxford University Press, New York, pp. 321–344.

Stumpf, R. M. & Boesch, C. (2006) The efficacy of female choice in chimpanzees of the Taï Forest, Cote d'Ivoire. *Behav Ecol Sociobiol* **60**: 749–765.

Sugiyama, Y. (1965) On the social change of hanuman langurs (*Presbytis entellus*). *Primates* **7**: 41–72.

Sugiyama, Y. (1981) Observations on the population dynamics and behavior of wild chimpanzees at Bossou, Guinea, in 1979–1980. *Primates* **22**: 435–444.

Sugiyama, Y. (1995) Tool-use for catching ants by chimpanzees at Bossou and Monts Nimba, West Africa. *Primates* **36**: 183–205.

Sugiyama, Y. (2009) Carrying of dead infants by Japanese macaque (*Macaca fuscata*) mothers. *Anthropol Sci* **117**: 113–119.

Sugiyama, Y. & Koman, J. (1987) A preliminary list of chimpanzees' alimentation at Bossou, Guinea. *Primates* **28**: 133–147.

Sussman, R. W. (2007) Brief history of primate field studies. In Campbell, C. J., Fuentes, A., MacKinnon, K. C., Panger, M. & Bearder, S. K. (eds), *Primates in Perspective*. Oxford University Press, New York, pp. 6–10.

Sutherland, S. (1987) Taste and smell. In McFarland, D. (ed.), *The Oxford Companion to Animal Behaviour*. Oxford University Press, Oxford, pp. 542–543.

Suzuki, A. (1969) An ecological study of chimpanzees in a savanna woodland. *Primates* **10**: 103–148.

Suzuki, A. (1971) Carnivory and cannibalism observed among forest-living chimpanzees. *J Anthropol Soc Nippon* **79**: 30–48.

Suzuki, S., Kuroda, S. & Nishihara, T. (1995) Tool-set for termite fishing by chimpanzees in the Ndoki Forest, Congo. *Behaviour* **132**: 219–235.

Symons, D. (1978) *Play and Aggression: A Study of Rhesus Monkeys*. Columbia University Press, New York.

Takahata, Y. (1985) Adult male chimpanzees kill and eat a male newborn infant: newly observed intragroup infanticide and cannibalism in Mahale National Park, Tanzania. *Folia Primatol* **44**: 161–170.

Takahata, H. & Takahata, Y. (1989) Inter-unit-group transfer of an immature male common chimpanzee and his social interactions in the non-natal group. *Afr Stud Monogr* **9**: 209–220.

Takahata, Y., Hasegawa, T. & Nishida, T. (1984) Chimpanzee predation in the Mahale Mountains from August 1979 to May 1982. *Int J Primatol* **5**: 213–233.

Takahata, Y., Hiraiwa-Hasegawa, M., Takasaki, H. & Nyundo, R. (1986) Newly acquired feeding habits among the chimpanzees of the Mahale Mountains National Park, Tanzania. *Hum Evol* **1**: 277–284.

Takahata, Y., Ihobe, H. & Idani, G. (1996) Comparing copulations of chimpanzees and bonobos: do females exhibit proceptivity or receptivity? In McGrew, W. C., Marchant, L. F. & Nishida, T. (eds), *Great Ape Societies*. Cambridge University Press, Cambridge, pp. 146–156.

Takasaki, H. (1983) Seed dispersal by chimpanzees: a preliminary note. *Afr Stud Monogr* **3**: 105–108.

Takasaki, H. & Hunt, K. (1987) Further medicinal plant consumption in wild chimpanzees? *Afr Stud Monogr* **8**: 125–128.

Takasaki, H. & Uehara, S. (1984) Seed dispersal by chimpanzees: supplementary note. *Afr Stud Monogr* **5**: 91–92.

Tanaka, J. (1976) Subsistence ecology of Central Kalahari San. In Lee, R. B. & DeVore, I. (eds), *Kalahari Hunter-Gatherers*. Harvard University Press, Cambridge, MA, pp. 98–119.

Tanzania National Parks (2006) *Mahale Mountains National Park: General Management Plan 2006–2016*. TANAPA, Arusha.

Teleki, G. (1973) *The Predatory Behavior of Wild Chimpanzees*. Bucknell University Press, Lewisbury, PA.

Teleki, G. (1989) Population status of wild chimpanzees (*Pan troglodytes*) and threats to survival. In Heltne, P. G. & Marquardt, L. A. (eds), *Understanding Chimpanzees*. Harvard University Press, Cambridge, MA, pp. 312–353.

Thornhill, R. (1976) Sexual selection and nuptial feeding behavior in *Bittacus apicalis* (Insecta: Mecoptera). *Am Nat* **110**: 529–548.

Thorpe, W. (1963) *Learning and Instinct in Animals*. Methuen, London.

Tinbergen, N. (1951) *The Study of Instinct*. Oxford University Press, Oxford.

Tomasello, M., Kruger, A. C., & Ratner H. H. (1993) Cultural learning. *Behav Brain Sci* **16**: 495–552.

Trivers, R. L. (1972) Parental investment and sexual selection. In Campbell, B. (ed.), *Sexual Selection and the Descent of Man 1871–1971*. Aldine, Chicago, pp. 136–179.

Trivers, R. L. (1974) Parent–offspring conflict. *Am Zool* **14**: 249–264.

Tsukahara, T. (1992) Lions eat chimpanzees: the first evidence of predation by lions on wild chimpanzees. *Am J Primatol* **29**: 1–11.

Turnbull, C. (1965) *Wayward Servants*. Natural History Press, New York.

Turner, L. A. (1995) Ntologi falls? *Pan Afr News* **2**(2): 9–10.

Turner, L. A. (1996) Invasive plant in chimpanzee habitat at Mahale. *Pan Afr News* **3**(1): 5.

Turner, L. A. (2006) Vegetation and chimpanzee ranging in the Mahale Mountains National Park, Tanzania. *Mem Fac Sci Kyoto Univ (Ser Biol)* **18**: 35–43.

Tutin, C. E. G. (1975) Exceptions to promiscuity in a feral chimpanzee community. In Kondo, S., Kawai, M. & Ehara, A. (eds), *Contemporary Primatology*. S. Karger, Basel, pp. 445–449.

Tutin, C. E. G. (1979a) Responses of chimpanzees to copulation, with special reference to interference by immature individuals. *Anim Behav* **27**: 845–854.

Tutin, C. E. G. (1979b) Mating patterns and reproductive strategies in a community of wild chimpanzees (*Pan troglodytes schweinfurthii*). *Behav Ecol Sociobiol* **6**: 29–38.

Tutin, C. E. G. & McGinnis, P. R. (1981) Chimpanzee reproduction in the wild. In Graham, C. E. (ed.), *Reproductive Biology of the Great Apes*. Academic Press, New York, pp. 239–264.

Uehara, S. (1982) Seasonal changes in the techniques employed by wild chimpanzees in the Mahale Mountains, Tanzania, to feed on termites (*Pseudacanthotermes spiniger*). *Folia Primatol* **37**: 44–76.

Uehara, S. (1986) Sex and group differences in feeding on animals by wild chimpanzees in the Mahale Mountains National Park, Tanzania. *Primates* **27**: 1–13.

Uehara, S. & Nyundo, R. (1983) One observed case of temporary adoption of an infant by unrelated nulliparous female among wild chimpanzees in the Mahale Mountains, Tanzania. *Primates* **24**: 456–466.

Uehara, S., Hiraiwa-Hasegawa, M., Hosaka, K. & Hamai, M. (1993) The fate of defeated alpha male chimpanzees in relation to their social networks. *Primates* **35**: 49–55.

de Waal, F. B. M. (1982) *Chimpanzee Politics*. Jonathan Cape, London.

de Waal F. B. M. (1989) *Peacemaking Among Primates*. Harvard University Press, Cambridge, MA.

de Waal, F. B. M. (1995) Bonobo sex and society. *Scient Amer* **272**: 58–64.

de Waal, F. B. M. (1998) *Chimpanzee Politics* (Revised Edition). Johns Hopkins University Press, Baltimore.

de Waal, F. B. M. (2003) Losing ten pounds to gain knowledge. *Pan Afr News* **10**: 17–19.

de Waal, F. B. M. & Lanting, F. (1997) *Bonobo*. University of California Press, Berkeley, CA.

Wakibara, J. V. (1998) Observations on the pilot control of *Senna spectabilis*, an invasive exotic tree in the Mahale Mountains National Park, Tanzania. *Pan Afr News* **5**: 4–6.

Waterman, P. G. (1984) Food acquisition and processing as a function of plant chemistry. In Chivers, D. J., Wood, B. A. & Bilsborough, A. (eds), *Food Acquisition and Processing in Primates*. Plenum, New York, pp. 177–211.

Watts, D. P. (2008) Tool use by chimpanzees at Ngogo, Kibale National Park, Uganda. *Int J Primatol* **29**: 83–94.

Whiten, A. (1999) Parental encouragement in *Gorilla* in comparative perspective: implications for social cognition and the evolution of teaching. In Parker, S. T., Mitchell, R. W. &, Miles, L. (eds), *The Mentalities of Gorillas and Orangutans*. Cambridge University Press, Cambridge, pp. 342–366.

Wilson, E. O. (1975) *Sociobiology*. Belknap Press, Cambridge, MA.

Wrangham, R. W. (1975) The behavioural ecology of chimpanzees in Gombe National Park, Tranzania. PhD thesis, University of Cambridge.

Wrangham, R. W. (1977) Feeding behavior of chimpanzees in Gombe National Park, Tanzania. In Clutton-Brock, T. H. (ed.), *Primate Ecology*. Academic Press, London, pp. 503–538.

Wrangham, R. W. & Nishida, T. (1983) *Aspilia* spp. leaves: a puzzle in the feeding behavior of wild chimpanzees. *Primates* **24**: 276–282.

Wrangham, R. W. & Riss, E. Z. B. (1990) Rates of predation on mammals by Gombe chimpanzees, 1972–1975. *Primates* **31**: 157–170.

Wrangham, R. W., Conklin-Brittain, N. L. & Hunt, K. D. (1998) Dietary response of chimpanzees and cercopithecines to seasonal variation in fruit abundance: I. Antifeedants. *Int J Primatol* **19**: 949–970.

Wrangham, R. W., Hagel, G., Leighton, M., *et al.* (2007) The great ape world heritage species project. In Stoinski, T. S., Steklis, H. D. & Mehlman, P. T. (eds), *Conservation in the 21st Century: Gorillas as a Case Study*. Springer, New York, pp. 282–295.

Yamagiwa, J. (1992) Functional analysis of social staring behavior in an all-male group of mountain gorillas. *Primates* **33**: 523–544.

Yamakoshi, G. (1998) Dietary responses to fruit scarcity of wild chimpanzees at Bossou, Guinea: possible implications for ecological importance of tool use. *Am J Phys Anthropol* **106**: 283–295.

Yerkes, R. M. (1943) *Chimpanzees: A Laboratory Colony*. Yale University Press, New Haven, CT.

Zamma, K. (2002) Leaf-grooming by a wild chimpanzee in Mahale. *Primates* **43**: 87–90.

Index

Printed in the United States
by Baker & Taylor Publisher Services